Social Equity in the Energy Sector

Indre Siksnelyte-Butkiene
• Dalia Streimikiene

Social Equity in the Energy Sector

palgrave
macmillan

Indre Siksnelyte-Butkiene
Kaunas Faculty, Vilnius University
Vilnius, Lithuania

Dalia Streimikiene
Kaunas Faculty, Vilnius University
Vilnius, Lithuania

ISBN 978-3-032-22223-7 ISBN 978-3-032-22224-4 (eBook)
https://doi.org/10.1007/978-3-032-22224-4

This Palgrave Macmillan imprint is published by the registered company Springer Nature Switzerland AG
The registered company address is: Gewerbestrasse 11, 6330 Cham, Switzerland

Contents

Figures

Tables

CHAPTER 1

Introduction

1.1 Why Social Equity Matters in Energy

Social equity constitutes a fundamental principle in the transformation of energy systems globally. As nations work toward achieving their climate goals, it has become increasingly evident that sustainable development must be closely aligned with social equity issues (Fragkos et al., 2021). The shift to low-carbon energy systems presents a significant opportunity to mitigate climate change by reducing greenhouse-gas (GHG) emissions. However, to fully realize the benefits of this transition, it is imperative to address the issue of social equity within energy policy in practice.

Energy is a critical enabler of economic and social activity, driving not only industry and commerce but also powering essential services in health, education, and mobility, as well as facilitating access to information and technology. The ability to access reliable and affordable energy is, therefore, a key determinant of social and economic inclusion. Despite this, millions of people—particularly those in marginalized communities—remain unable to access the benefits of modern energy systems. Energy poverty, defined as the lack of access to affordable and reliable energy, disproportionately affects vulnerable populations, including low-income households, women and the elderly, and rural communities. This lack of access exacerbates existing social inequalities, limiting opportunities for economic advancement, improved health outcomes, and social integration.

The principle of social equity in energy systems is closely tied to human rights and social justice. Access to basic energy services is not simply a matter of convenience; it is a fundamental human right. The United Nations' Sustainable Development Goal 7 (SDG 7) (The United Nations, 2015a) highlights the necessity of ensuring universal access to affordable, reliable, sustainable, and modern energy. Additionally, SDG 10 calls for reducing inequalities both within and among countries. However, the realization of

I. Siksnelyte-Butkiene, D. Streimikiene, *Social Equity in the Energy Sector*, https://doi.org/10.1007/978-3-032-22224-4_1

these goals requires a systemic approach to addressing disparities in energy access, which are often rooted in broader socioeconomic inequalities. Vulnerable groups, such as rural populations, women, children, the elderly, and indigenous communities, bear a disproportionate burden when it comes to energy-related challenges. If these inequities remain unaddressed, energy systems will perpetuate cycles of poverty and exclusion, reinforcing socioeconomic disadvantages across generations.

The global energy transition, driven by the need to combat climate change and meet the Paris Agreement targets (The United Nations, 2015b), presents both significant opportunities and risks from an equity perspective. On one hand, a shift towards renewable energy sources, such as solar, wind, and hydropower, holds the potential to offer cleaner, more sustainable energy solutions that can be more widely accessible. On the other hand, the transition also presents challenges in terms of ensuring that low-income households, women, the elderly, and other vulnerable communities are not left behind. The costs of renewable technologies, the loss of jobs in traditional energy sectors, and the digital divide in energy access can deepen inequalities if not properly managed.

In addition to access and affordability, social equity in energy systems must also incorporate broader societal dimensions. These include health impacts from air pollution and the potential for renewable energy expansion to reduce public health burdens. Furthermore, the energy transition holds promise for employment generation in clean energy sectors, but targeted policies are necessary to ensure that job creation is inclusive and supports communities affected by the decline of fossil-fuel industries.

Furthermore, as energy markets evolve with the introduction of new technologies, such as smart grids, energy storage, and electric vehicles, there is a growing concern that the benefits of these innovations will be concentrated in wealthier regions and among more affluent populations, while marginalized communities may be excluded. Technological innovations in energy efficiency, storage, and smart metering must be critically evaluated for their potential to either promote or undermine social equity. Policy frameworks should be designed to ensure that these technological advances contribute to a just transition, making certain that the benefits of modern energy systems reach all sectors of society.

The particular challenges faced by elderly populations and women must also be addressed in the context of energy transitions. Older adults may encounter difficulties in adopting new energy technologies due to physical limitations, while women often bear a disproportionate responsibility for managing energy consumption at the household level. Similarly, marginalized communities often lack the necessary resources to transition to more sustainable energy systems, leaving them vulnerable to the social and economic effects of an energy system that does not cater to their specific needs.

In this context, social equity in energy systems is essential not only for poverty alleviation and human development but also for the long-term sustainability of energy systems. Energy systems that fail to incorporate equity considerations will face not only social resistance but also economic inefficiencies and political challenges. A just energy transition, which prioritizes equity alongside sustainability, can contribute to greater social resilience, political stability, and the creation of more inclusive societies. Consequently, social equity should be seen not as an optional component but as a core element in the design of energy systems and the formulation of energy policies.

Ultimately, social equity in energy systems is crucial to addressing the interconnected challenges of climate change, poverty, and inequality. Energy systems that do not provide universal access to affordable, clean energy undermine efforts to achieve broader sustainable development goals. By embedding equity into energy policy, governments, international organizations, and energy providers can ensure that the benefits of energy transitions are shared equitably across all segments of society, particularly those facing additional barriers due to age, gender, or socioeconomic status. This monograph investigates the importance of social equity in energy systems and seeks to foster the development of more inclusive, resilient, and sustainable energy frameworks that promote fairness, justice, and long-term social stability.

1.2 Aim of the Monograph

The primary aim of this monograph is to critically examine how social equity principles can be embedded within energy systems and policies, ensuring that the ongoing global energy transition is both environmentally sustainable and socially just. As energy systems transition towards more sustainable, low-carbon models, the monograph emphasizes that ensuring equitable access to energy services is not merely a desirable goal but a critical component of social justice. It highlights how different societal groups are often disproportionately affected by energy transitions. This monograph seeks to pay attention to the central challenge of fostering inclusivity within the context of global energy transitions, ensuring that the benefits of cleaner, more sustainable energy are shared equitably across all segments of society.

The monograph further aims to advance methodological and policy pathways for measuring, assessing, and operationalizing social equity in the energy sector. In this context, the monograph underscores the need for inclusive and gender-sensitive policies, age-inclusive measures, and policies that tackle energy inequity and ensure universal access to affordable, reliable, and clean energy for all.

The monograph seeks to provide a comprehensive and multidimensional perspective on social equity in the energy sector. By exploring the key challenges, policy solutions, and technological innovations necessary for equitable energy transitions, it contributes valuable insights for policymakers, researchers, and energy practitioners.

1.3 Methodological Approach

The research methodology employed in this monograph utilizes a mixed-methods approach. However, a comprehensive analysis of scientific literature serves as a foundation for identifying the main theories, concepts, and debates surrounding the issues of social equity in the energy sector. The literature review provides essential insights into the theoretical framework and ongoing discussions. Following this, governmental and policy documents are examined to assess how current policy frameworks either address or neglect social equity concerns. These documents shed light on the priorities and objectives of policymakers concerning inclusive energy transitions. Additionally, empirical studies are incorporated to capture the diverse experiences of marginalized groups across different geographical regions, with a particular focus on the impacts of energy policies on vulnerable populations such as low-income households, women, and the elderly. Analysis of different case studies provides a more practical perspective on how energy equity initiatives are implemented in different contexts.

1.4 Scope and Structure of the Book

The scope of this monograph encompasses the conceptual foundations, analytical frameworks, and policy applications necessary for understanding and promoting social equity within energy systems. It situates the discussion within broader global and European policy contexts, linking energy equity and justice debates with the demographic shifts, the challenges of sustainable development, energy decarbonization, and digitalization. The analysis in this monograph unfolds through seven interconnected chapters, each addressing a key component of the equity–energy nexus.

Chapter 2 lays the conceptual and theoretical foundation by defining social equity and energy justice and identifying their main dimensions. It further explores the nexus between these two concepts, providing a framework that guides the analysis throughout the book.

Chapter 3 focuses on the intersection between energy transition and social justice. It introduces the just transition framework and examines the distributional consequences of decarbonization processes. Special attention is given to intergenerational equity, emphasizing how long-term sustainability requires balancing present and future welfare outcomes.

Chapter 4 explores the relationship between energy poverty and social equity. It reviews the conceptual evolution of energy poverty, presents different measurement approaches, and discusses how social equity considerations can enhance the assessment and mitigation of energy poverty across diverse socioeconomic contexts.

Chapter 5 investigates the gendered aspects of energy systems, addressing the vulnerabilities and opportunities faced by women in the context of energy transitions. It reviews frameworks and empirical evidence on gender mainstreaming, energy access, and household decision-making, emphasizing the importance of gender-responsive policies for achieving inclusive energy transitions.

Chapter 6 turns to age-based equity, a dimension often overlooked in energy studies. It introduces the concept of energy ageism and explores how aging populations experience inequities in access, affordability, and digital engagement with new energy technologies. Comparative and international evidence is presented to illustrate the social, spatial, and policy determinants of age-related disparities.

Chapter 7 synthesizes analytical and policy approaches to advancing social equity in the energy sector. It reviews existing metrics, indicators, and methodologies for measuring social equity, alongside regulatory, financial, and technological instruments designed to enhance fairness in energy systems. The chapter identifies persistent data and measurement gaps and proposes ways to improve policy integration.

Chapter 8 proposes a social equity framework for an age-inclusive low-carbon transition. It examines governance models, institutional mechanisms, and welfare approaches that can operationalize just transition principles in practice. The chapter concludes with reflections on future research directions and the challenges of implementing socially equitable energy policies in an era of rapid technological and environmental change.

The Conclusion synthesizes the central findings of the monograph, highlighting that embedding social equity within the energy transition is not only a moral and social imperative but also a prerequisite for ensuring political legitimacy, economic efficiency, and environmental sustainability.

This monograph is intended for a broad readership that includes academics, policymakers, and practitioners engaged in energy, sustainability, and social policy research.

References

Fragkos, P.; Fragkiadakis, K.; Sovacool, B.; Paroussos, L.; Vrontisi, Z.; Charalampidis, I. Equity implications of climate policy: Assessing the social and distributional impacts of emission reduction targets in the European Union. Energy, 2021, 237, 121591.

The United Nations. Transforming our world: the 2030 Agenda for Sustainable Development. Resolution adopted by the General Assembly on 25 September 2015, 2015a. Available online: https://docs.un.org/en/A/RES/70/1 [assessed on 02 June 2025].

The United Nations. Paris Agreement. 2015b. Available online: https://unfccc.int/sites/default/files/nglish_paris_agreement.pdf [assessed on 02 June 2025].

CHAPTER 2

Understanding Social Equity in the Energy Sector

2.1 Defining Social Equity and Energy Justice

Ensuring fairness in the global shift toward sustainable energy systems has become an essential concern in energy policy and research. The pursuit of climate-neutral economies, while technologically achievable, raises complex socioeconomic challenges, especially in terms of how energy policies affect different social groups. Two interrelated but distinct concepts—energy justice and social equity in the energy sector—have emerged to address these challenges. While both frameworks emphasize fairness and inclusion, their theoretical roots, analytical focus, and application in energy governance differ.

2.1.1 Defining Energy Justice

Amid the intensifying global pursuit of decarbonization and climate neutrality, the concept of energy justice has gained increasing prominence as a critical analytical and normative lens for assessing the fairness of energy systems. The accelerating pace of the low-carbon transition has brought into sharp focus the persistent inequalities in energy access, affordability, and governance—revealing that technological innovation, while essential, is insufficient to address deeply embedded structural disparities. This recognition has prompted a growing application of justice principles to energy policy, underscoring the need for frameworks that evaluate not only efficiency and innovation but also equity and inclusion.

Energy justice offers a multidimensional approach to understanding how energy-related benefits and burdens are distributed across diverse social groups and geographical regions. It addresses the mechanisms through which decisions are made, the voices that are heard or excluded, and the societal norms and institutional structures that sustain or challenge inequitable outcomes. Rooted

I. Siksnelyte-Butkiene, D. Streimikiene, *Social Equity in the Energy Sector*,
https://doi.org/10.1007/978-3-032-22224-4_2

in a rich tradition of justice-oriented thought—drawing from social justice, environmental justice, political philosophy, and human rights discourses—energy justice serves as a conceptual bridge that links normative ideals with the realities of energy governance and practice.

In their comprehensive review of the energy justice literature, Jenkins et al. (2021) identified a wide array of classifications reflecting the diversity of marginalized groups and individuals disproportionately affected by injustices within energy systems. A total of 34 distinct coding categories were discerned, capturing a broad spectrum of vulnerabilities. These included demographic variables such as age, health status, disability, and caregiving responsibilities, alongside material conditions—such as residency in substandard housing—as well as procedural and structural exclusions, including limited participation in decision-making processes or opposition to dominant sociotechnical transformations. The analysis further encompassed contextual vulnerabilities, notably those arising from conflict, corruption, or labor displacement due to fossil-fuel sector decline. In several instances, categories aggregated overlapping forms of disadvantage, such as 'poor and economically marginalized', which encompassed financial, energy, and fuel poverty. Similarly, 'socially marginalized' referred to individuals constrained by caste, class, literacy, or language barriers. Collectively, these findings underscore the intricate interplay between social, material, and institutional factors in shaping energy-related vulnerabilities and highlight the necessity for justice frameworks sensitive to intersectional inequities.

The concept of energy justice proposes a systematic approach to restructuring energy systems in ways that are more socially responsive and ethically grounded. By interrogating who is impacted, how, and through what institutional channels, energy justice provides a platform for ensuring that no group is disproportionately burdened or left behind in the pursuit of sustainability. As energy transitions reshape socioeconomic landscapes, the imperative to integrate justice considerations becomes not only a moral obligation but a precondition for legitimacy and long-term resilience in energy governance.

Energy justice encompasses several key dimensions that have become central in the scientific discourse on sustainable energy (Heffron, 2023; Fig. 2.1).

The first dimension, ***distributive justice***, is concerned with the fair allocation of both the positive outcomes (such as access to clean and affordable energy) and the negative externalities (such as environmental pollution or displacement from energy infrastructure) associated with energy production and consumption. In practice, distributive justice is often compromised in low-income and rural areas, where structural barriers persist in accessing energy infrastructure, thus reinforcing patterns of energy poverty and social exclusion.

The second component, ***procedural justice***, emphasizes the importance of transparent, inclusive, and participatory governance in energy systems. This dimension focuses on ensuring that all stakeholders—particularly vulnerable and historically marginalized communities—are meaningfully involved in energy planning and decision-making. A failure to ensure procedural justice

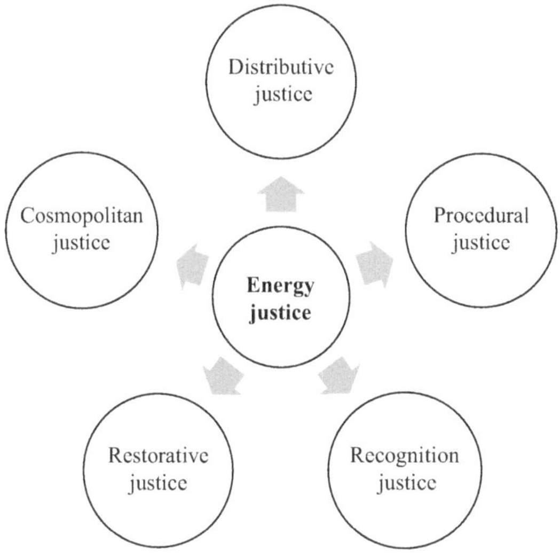

Fig. 2.1 The main dimensions of energy justice. Source: created by the author.

often results in top-down decisions that lack local legitimacy, create public opposition to energy projects, and undermine long-term policy sustainability.

A third fundamental pillar of energy justice is *recognition justice*, which refers to the acknowledgement and respect of cultural identities, socioeconomic differences, and historical inequalities that shape people's experiences of energy systems. Recognition justice requires that diverse perspectives, especially those of disadvantaged groups, be integrated into energy policy design. Without this recognition, policy measures risk reproducing existing disparities or introducing new forms of exclusion.

Recent literature has also highlighted additional dimensions of energy justice that are gaining significance in international energy policy. For example, *restorative justice* involves the need to address and rectify past harms caused by unjust energy policies and practices, particularly in regions that have experienced environmental degradation or economic disempowerment as a result of fossil-fuel exploitation. Furthermore, *cosmopolitan justice* underscores the global nature of energy-related externalities and responsibilities, emphasizing the ethical obligation of affluent societies to support energy access and decarbonization in the Global South.

In addition to these core and emerging dimensions, recent scholarship has introduced the concept of *energy sovereignty* as a complementary yet increasingly salient element within the energy justice discourse. Energy sovereignty refers to the right of individuals, communities, and nations to define, control, and benefit from their own energy systems in accordance with their social, cultural, and ecological values. In many analytical contexts, it is viewed as a cross-cutting principle that reinforces procedural and recognition justice by emphasizing agency, ownership, and self-determination in energy governance. However, in other contexts—particularly those examining community energy, indigenous self-governance, or local energy democracy—energy sovereignty is conceptualized as a distinct dimension of energy justice, reflecting the growing importance of autonomy and empowerment in just energy transitions. This perspective underscores that fairness in energy systems requires not only participation but also the capacity of communities to exercise meaningful control over their energy futures.

The integration of these dimensions into policy design and evaluation reflects a broader shift towards normative frameworks that prioritize not only efficiency and cost-effectiveness but also the ethical foundations of energy governance. For example, in the European context, particularly under the European Green Deal and the Just Transition Mechanism, the application of energy justice principles has become central to ensuring that climate neutrality is achieved without leaving behind vulnerable regions and populations.

From a policy perspective, operationalizing energy justice requires institutional mechanisms that address distributional impacts, promote participatory governance, and ensure accountability in energy decision-making. This includes targeted support for low-income households, enhanced stakeholder consultation processes, and recognition of local knowledge in the deployment of renewable energy technologies.

In summary, energy justice provides a robust conceptual basis for integrating ethical considerations into energy policy and for evaluating the socioeconomic impacts of decarbonization strategies. Its multidimensional framework—encompassing distributive, procedural, recognition, restorative, and cosmopolitan justice—enables policymakers to address the complexity of sustainable energy transitions in a manner that is both socially inclusive and environmentally sustainable. Therefore, the application of energy justice is not only desirable but necessary for achieving a just and equitable development of energy systems.

2.1.1.1 Energy Justice and Its Relationship to Social Justice

Energy justice is fundamentally embedded within the broader project of social justice, serving as a focused articulation of justice principles specifically adapted to the complexities of energy systems. While social justice encompasses the equitable distribution of resources, rights, and opportunities across all societal domains, energy justice translates these imperatives into the context of energy governance, access, and sustainability. This contextualization is necessary because energy systems intersect with unique technical, institutional, and sociopolitical challenges that demand tailored normative and analytical frameworks.

The relationship between energy justice and social justice is thus one of specificity within generality: energy justice represents the domain-specific expression of social justice ideals, illuminating how systemic inequalities manifest and can be addressed in energy-related issues. This linkage ensures that energy justice is not an isolated concept but an integral component of the broader social justice discourse, reinforcing ethical commitments to fairness, recognition, participation, and redress.

Moreover, energy justice enriches social justice debates by foregrounding the material realities and institutional mechanisms through which energy-related disparities occur. It compels attention to the distributive consequences of energy production and consumption, the inclusivity of decision-making processes, and the recognition of diverse community identities and histories—elements that are equally central to social justice but gain particular urgency in energy contexts. This focus is crucial as energy transitions have profound implications for economic opportunity, environmental health, and social cohesion, especially for historically marginalized and vulnerable populations.

In essence, energy justice operationalizes social justice within the energy sector by providing concrete criteria and frameworks that policymakers, researchers, and practitioners can employ to evaluate and promote fair and inclusive energy futures. Recognizing this intrinsic relationship underscores the ethical necessity of integrating social justice principles into energy policies and interventions, ensuring that transitions toward sustainability are both just and equitable at a societal scale.

2.1.2 Defining Social Equity

The development of the energy sector has profound implications for social inclusion, public welfare, and the equitable distribution of essential services. As a consequence, the concept of social equity is increasingly recognized as a critical policy objective, requiring governments and energy institutions to ensure that all social groups—particularly vulnerable and disadvantaged populations—have fair access to energy services and are able to benefit from energy-related development. Social equity refers not only to formal equality in access to infrastructure, but also to substantive fairness in how energy services are distributed, governed, and experienced across different socioeconomic contexts.

Empirical evidence highlights the broad and interconnected consequences of social inequity, extending to domains such as health, subjective well-being, environmental attitudes, and institutional trust, among others. It is empirically proved that societies with higher levels of income inequality tend to experience more pronounced social problems and reduced population health outcomes. Similarly, it is also found that greater income concentration correlates with lower individual life satisfaction and more frequent negative emotional states. Further studies suggest that nations with higher income equality exhibit stronger pro-environmental behaviors, such as elevated recycling rates, potentially reflecting more widespread environmental concern. In addition, higher levels of opportunity and income equality have been associated with stronger social trust and lower levels of perceived corruption, indicating the broader societal benefits of reducing structural inequalities (Pellegrini-Masini et al., 2020).

A foundational reference for conceptualizing social equity is found within the energy justice literature (Kime et al., 2023), where the dimensions of social equity are frequently aligned with those of energy justice, reflecting their close conceptual interrelationship. However, it is important to distinguish between the two concepts. In this book, the dimensions of energy justice and social equity are intentionally differentiated to reflect their distinct analytical purposes. Energy justice dimensions represent the normative and ethical foundations of fairness in energy systems. They provide the philosophical lens through which inequalities, responsibilities, and rights are interpreted and evaluated. In contrast, the dimensions of social equity serve an operational function, translating normative justice principles into measurable and policy-relevant outcomes that directly affect households, communities, and regions. This distinction is particularly important for empirical and policy-oriented research, as it enables a focus on tangible aspects of equity without conflating them with broader ethical or global justice frameworks. Accordingly, while the subsequent chapters of this book frequently reference justice dimensions to discuss the theoretical underpinnings of fairness, the equity dimensions are used when assessing how those justice principles manifest in practice.

While energy justice addresses the underlying questions of why and how fairness should be embedded in energy governance, social equity focuses on achieving fair and inclusive outcomes through targeted interventions. This pragmatic orientation involves addressing critical factors such as the availability, affordability, reliability, and participatory inclusion in energy access. Consequently, the principal dimensions of social equity in the energy sector can be structured across these four domains—access, affordability, reliability, and participation—thereby offering a coherent framework for assessing and advancing equitable energy systems (Fig. 2.2). This distinction underscores that social equity, while informed by justice principles, emphasizes measurable improvements in reducing inequities faced by marginalized and vulnerable populations, thereby ensuring that energy transitions are not only fair in theory but also equitable in practice.

On the one hand, access to energy services refers to the ability of individuals and communities to obtain reliable, sufficient, and safe energy for essential needs such as heating, lighting, cooking, communication, and mobility. Reliability, in this context, emphasizes the consistent and uninterrupted availability of energy, which is crucial for maintaining daily life functions and preventing additional costs or hardships associated with outages and unstable supply. Inequities in reliability often compound access issues, as vulnerable populations tend to experience more frequent and prolonged disruptions, limiting their capacity to depend on energy services for health, education, and economic activities. Inequities in access often emerge due to geographic isolation (e.g., in rural or remote areas without grid connections), economic constraints (e.g., households unable to afford rising energy prices), or institutional limitations (e.g., exclusion from subsidy programs). These barriers result in energy deprivation, which can in turn reinforce broader forms of social and economic inequality, including educational disadvantages, health risks, and restricted employment opportunities.

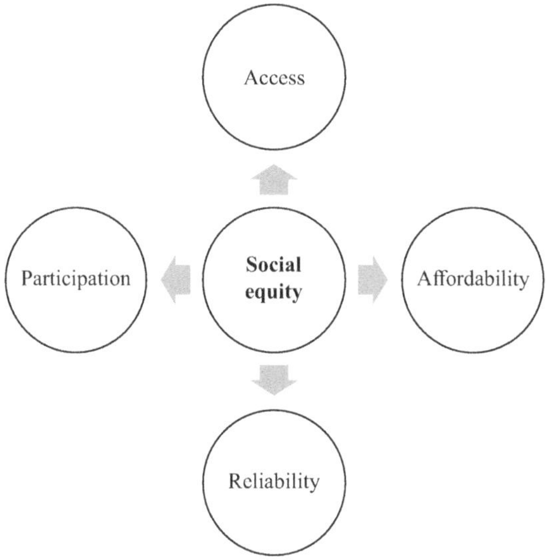

Fig. 2.2 The main dimensions of social equity in the energy sector. Source: created by the author.

On the other hand, disparities in energy outcomes refer to the unequal distribution of the benefits and burdens of energy sector development. While certain groups may gain from energy investments through improved reliability, employment, or enhanced infrastructure, other communities may experience adverse effects such as environmental degradation, land expropriation, or exposure to pollution. Social equity requires that these outcomes be distributed in a fair manner so that the gains from energy development are shared widely and no population group is disproportionately exposed to risks or harms. This dual emphasis on access and outcomes provides a more comprehensive understanding of equity than only distribution.

Another central component of social equity is affordability. Rising energy prices, deregulated energy markets, and declining public subsidies can significantly affect energy affordability for lower-income groups. In this regard, energy affordability is not only an economic concern but also a social one, as inadequate access to heating, electricity, or transportation energy can lead to broader forms of deprivation, including energy poverty. As energy represents a basic human need and a precondition for social participation, unaffordable energy threatens both welfare and dignity. In this context, policies such as progressive pricing schemes, targeted subsidies, and energy poverty mitigation programs are crucial to uphold equity principles in energy pricing.

Geographical equity is also a key concern, particularly in countries with different regional development level. Disparities in energy infrastructure investment between urban and rural areas may lead to persistent territorial inequalities. For example, regions that lack centralized heating networks or suffer from inadequate electricity infrastructure, and limited availability of modern fuels face higher costs and lower quality of service, thereby limiting local development. Ensuring social equity requires that energy infrastructure expansion be guided not only by market logic but also by spatial justice considerations, including the public interest in regional cohesion and balanced development.

Participation and voice also constitute important elements of social equity, particularly in relation to how local communities are engaged in the planning and implementation of energy projects. While this overlaps with procedural justice, in the equity context it is particularly relevant to ensure that disadvantaged or underrepresented groups have opportunities to influence decisions that affect their living conditions. Public consultations, community-energy initiatives, and social dialogue mechanisms can enhance the equity outcomes of energy infrastructure development and foster local ownership.

Social equity is also increasingly relevant in the context of institutional design. Regulatory frameworks, pricing mechanisms, and market reforms must be carefully evaluated for their social impacts. For instance, liberalization and privatization processes in the energy sector may improve market efficiency but could also exacerbate inequality if not accompanied by adequate regulatory safeguards. In this respect, social-impact assessments and equity audits should become standard practice in energy sector governance.

In addition, the role of education and energy literacy is critical in advancing social equity. Low awareness and technical knowledge among vulnerable populations can act as a barrier to accessing support programs, adopting new technologies, or participating in demand-side energy initiatives. Policymakers must therefore implement outreach programs, public awareness campaigns, and tailored communication strategies to ensure inclusiveness in energy policy implementation.

Finally, the implementation of socially equitable energy policies requires continuous data collection, monitoring, and evaluation. Without disaggregated data on energy access, affordability, and quality, policymakers lack the evidence base needed to design and adjust interventions. Equity audits, social-impact assessments, and participatory monitoring tools are vital to ensure that energy sector development aligns with social equity objectives.

Therefore, it can be stated that social equity in the energy sector is a multidimensional policy objective that addresses economic, geographic, and institutional barriers to fair energy access. It complements the energy justice framework by focusing on the distributional and participatory aspects of energy development, particularly for those groups most at risk of exclusion. As energy systems evolve and diversify, ensuring social equity will be essential for maintaining public support, enhancing legitimacy of policy decisions, and fostering socially inclusive energy development.

2.1.2.1 Social Equity as a Dimension of Social Justice

It is important to recognize that social equity represents a critical dimension within the broader framework of social justice. While social justice encompasses comprehensive ethical principles aimed at achieving fairness across distribution, recognition, participation, and institutional structures, social equity specifically addresses the tangible and measurable aspects of fairness related to access, affordability, reliability, and participation within the energy sector.

In this context, social equity can be understood as an operational subset of social justice: it translates the normative ideals of justice into concrete policies and practices that reduce disparities and promote fairness in energy access and outcomes. Social equity focuses on identifying and addressing the specific barriers and inequalities faced by marginalized and vulnerable groups, thereby contributing to the realization of social justice in practical terms.

Thus, social equity is not separate from social justice but embedded within it as a necessary and actionable component. It serves as the foundation upon which social justice principles are implemented and evaluated in the energy sector. Recognizing social equity as part of social justice highlights the importance of both principled commitments and pragmatic interventions in ensuring fair and inclusive energy systems.

2.2 Main Dimensions of Energy Justice

The energy justice concept has emerged as a comprehensive analytical and normative lens through which the equity dimensions of contemporary energy systems are examined. It addresses the growing imperative to integrate ethical considerations into energy governance, particularly in light of ongoing low-carbon transitions that frequently exacerbate preexisting socioeconomic, geographical, and intergenerational disparities. Rooted in broader discourses of social and environmental justice, as well as political philosophy, energy justice offers a means to assess and guide the distribution of benefits and burdens across diverse societal groups. It thereby repositions energy policy not solely as a technical or economic issue but as a fundamentally social challenge requiring fair and inclusive solutions. The newest literature generally conceptualizes energy justice as comprising three traditional dimensions—distributional, recognition, and procedural justice—as well as two emerging ones: restorative and cosmopolitan justice.

2.2.1 *Distributional Justice*

Distributional justice within energy systems addresses the unequal allocation of both benefits and burdens associated with energy production, consumption, and infrastructure development. This dimension recognizes that environmental advantages—such as clean air, reliable energy access, and economic opportunities—and environmental harms—such as pollution, displacement, and health risks—are rarely shared equitably across society. Importantly, distributional justice also extends to the uneven assignment of responsibilities and obligations related to energy transitions, such as who bears the costs of mitigation measures or infrastructure siting.

A central concern of distributional justice is the spatial and social disparities that emerge from the physical placement of energy infrastructure. The geographic distribution of resources, like wind or solar potential, is inherently uneven, which complicates efforts to achieve equitable outcomes. For example, communities located near fossil-fuel extraction sites or renewable energy installations often experience disproportionate environmental and social impacts, raising critical questions about fairness in both the selection of sites and the compensation or protections provided to affected populations. Empirical analyses thus focus on assessing how infrastructure siting contributes to distributional inequities, highlighting cases where marginalized groups bear a disproportionate share of risks and adverse effects.

Beyond infrastructure, distributional justice also encompasses access to energy services. The issue of energy affordability—commonly framed within the fuel poverty discourse—exposes the socioeconomic dimensions of inequity, demonstrating how low-income and vulnerable households frequently face barriers to securing reliable and affordable energy for basic needs like heating, cooling, and lighting (Jenkins et al., 2016; Tarekegne, 2020). This lack of access not only threatens physical well-being but also restricts individuals' capabilities and freedoms, limiting their ability to fully participate in social and economic life. Consequently, distributional justice calls for policies that ensure equitable availability and affordability of energy services, recognizing energy as a fundamental necessity rather than a commodity accessible only to those with sufficient means.

Moreover, distributional justice primarily concerns the equitable allocation of benefits and burdens among current social groups during the energy transition. This includes addressing how costs and advantages—such as job displacement in fossil-fuel sectors or changes in energy prices—are fairly shared and mitigated through compensatory measures and social protections. Closely related, intergenerational equity emphasizes the responsibility to ensure that present energy policies and practices do not impose disproportionate risks or deprivations on future generations. Together, these dimensions underscore the necessity of a just transition that accounts for fairness both within and across generations.

It can be stated that distributional justice requires a comprehensive approach that integrates concerns about the uneven geographic and social distribution of energy-related benefits and burdens, the accessibility and affordability of energy services, and the equitable sharing of responsibilities across society. It calls for a shift from a narrow focus on technical and economic efficiency toward a broader ethical commitment to fairness, inclusivity, and social welfare in energy systems.

2.2.2 Recognition Justice

Recognition justice stands as a core pillar within the energy justice framework, emphasizing the vital need to acknowledge, respect, and validate the identities, rights, and lived experiences of all social groups impacted by energy systems.

It goes beyond questions of fair distribution or inclusive participation to critically examine the cultural and institutional structures that render certain populations invisible, misrepresented, or devalued within energy policy, planning, and governance.

At its essence, recognition justice demands the dismantling of systemic marginalization manifested through cultural domination or exclusion. It insists that vulnerable and historically disenfranchised groups—such as indigenous peoples, ethnic minorities, low-income communities, and gender-diverse populations—not only be included in energy discussions but also receive full recognition of their sociocultural contexts, knowledge systems, and political agency. Achieving this requires a fundamental transformation of institutional practices, which have long privileged dominant perspectives while overlooking or misinterpreting the needs and contributions of marginalized actors.

Fraser (1995) identified three primary forms of misrecognition: cultural domination, where dominant norms suppress minority identities; non-recognition, where entire groups are excluded from decision-making; and disrespect, where individuals or communities are stigmatized, stereotyped, or denied rights.

Recognition justice also interrogates how power shapes the definition of expertise and determines whose knowledge is considered legitimate. In energy policymaking, technical and economic rationales frequently prevail, marginalizing other forms of understanding—ethical, cultural, place-based, or experiential. This epistemic injustice further entrenches exclusion, as the communities most affected by energy infrastructure are seldom recognized as important contributors to policy deliberation or innovation.

Crucially, recognition justice requires institutional reform. Energy governance frameworks must be restructured to embrace pluralistic worldviews, particularly those grounded in local traditions, indigenous knowledge systems, and gender-sensitive perspectives. This involves redesigning participatory processes to move beyond mere formality, ensuring they are genuinely inclusive and culturally respectful. Central to recognition justice is the right of communities not only to be heard but to actively shape policy agendas.

Finally, recognition justice is deeply intertwined with distributional and procedural justice. Without genuine recognition, participatory efforts risk becoming tokenistic, and distributive outcomes may perpetuate historical injustices. Recognition is both a prerequisite and companion to other dimensions of energy justice—it underpins equitable access, enhances the legitimacy of decision-making, and ensures that energy transitions reflect the diverse values and aspirations of affected populations.

2.2.3 Procedural Justice

Procedural justice represents a key dimension of the energy justice framework, focusing on the fairness and transparency of decision-making processes related to energy governance. While distributive and recognition justice address the outcomes and identities involved, procedural justice is concerned with how decisions are made, who is allowed to participate, and to what extent their voices influence outcomes.

At its core, procedural justice emphasizes inclusive, transparent, and accountable processes that ensure all affected stakeholders—particularly those from marginalized or vulnerable groups—have a meaningful opportunity to participate in shaping energy policy and planning. It challenges institutional structures that centralize power and limit access to information, decision-making forums, or mechanisms for redress.

Key elements of procedural justice include early and continuous stakeholder engagement, the right to access relevant information in understandable formats, and the assurance that public input can meaningfully influence outcomes (Korsnes and Throndsen, 2021). Participation should not be limited only to formal consultation, which often has limited impact. Instead, procedural justice calls for deliberative and collaborative approaches that value diverse knowledge systems, community priorities, and place-based experiences. Given that participation lies at the core of procedural justice and is intended to empower individuals in decision-making processes (Simcock, 2016) it is essential that participatory processes also incorporate structured community engagement measures. Empirical evidence presented by Oliveira et al. (2025) underscores the significance of governmental support measures in fostering public participation, highlighting the role of institutional incentives in promoting active community involvement.

Procedural injustices are particularly evident in energy infrastructure development, where communities—especially those already disadvantaged—are frequently excluded from planning processes or involved only after key decisions have been made. This exclusion can lead to mistrust, social conflict, and resistance, especially when people perceive decisions to be imposed without adequate consultation or consent (Ryder et al., 2023; Campos et al., 2025).

In the context of the energy sector, procedural justice is underpinned by principles that emphasize fairness, inclusivity, and transparency in decision-making processes that impact individuals and communities. These principles aim to ensure that stakeholders are not only adequately informed but also have opportunities to participate in decision-making. The literature presents various categorizations of procedural justice principles. For instance, one of the most detailed categorizations is provided by Ruano-Chamorro et al. (2022), where the authors contributed to a nuanced understanding of procedural justice in energy governance, proposing a structured framework that delineates eleven procedural justice criteria grouped under three overarching domains: process properties, participant agency, and interpersonal treatment (Fig. 2.3).

Process properties define the foundational conditions necessary for fair procedures and include transparency, accountability, neutrality, correctability, ethicality, and trustworthiness. The domain of agency highlights the capacity of individuals to engage autonomously in decision-making and comprises voice, decision control, and capabilities. Finally, interpersonal treatment reflects the quality of social

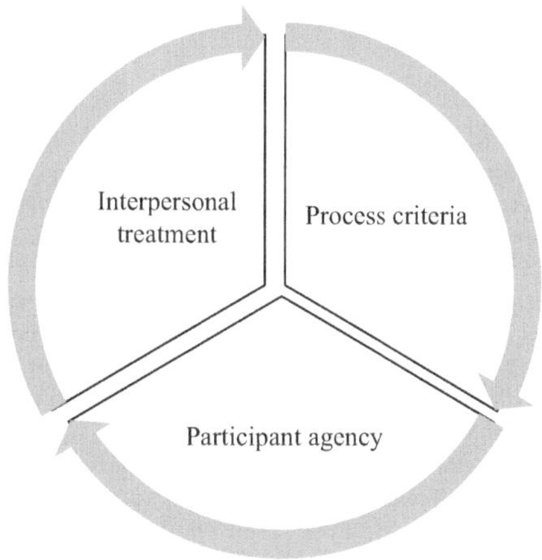

Fig. 2.3 Domains of procedural justice. Source: developed by the author based on Ruano-Chamorro et al. (2022).

interaction within participatory settings, defined through the criteria of respect and politeness. Table 2.1 overviews these principles one by one.

Moreover, procedural justice is closely linked to broader issues of political representation and institutional accountability. It requires governments and energy institutions to build decision-making frameworks that are not only legally inclusive but also culturally sensitive, linguistically accessible, and responsive to the needs of diverse groups. Special attention must be paid to overcoming barriers faced by underrepresented populations, including linguistic minorities, those with limited literacy, or communities with historically limited access to political processes.

Procedural justice also intersects with recognition justice. Without acknowledging and respecting the identities and experiences of different social groups, participatory mechanisms risk becoming superficial or tokenistic. In this sense, meaningful participation is both a right and a recognition of the agency and capacity of all individuals to influence decisions that affect their lives.

In the context of energy transitions, procedural justice becomes especially important. Decisions around the deployment of renewable technologies, the retirement of fossil-fuel infrastructure, or the allocation of climate finance have far-reaching implications. Ensuring that these decisions are made through fair and inclusive procedures strengthens social legitimacy, enhances public trust, and improves the sustainability of policy outcomes (Ellis et al., 2023).

Ultimately, procedural justice calls for more than technical adjustments to governance systems—it requires a shift in institutional culture toward openness, responsiveness, and respect for pluralism. By enabling meaningful participation and ensuring accountability, procedural justice helps create energy systems that are not only efficient and low-carbon but also socially just and democratically legitimate.

Table 2.1 Principles of procedural justice.

Principle	*Explanation*
Transparency	Refers to the open and honest dissemination of all relevant information throughout the decision-making process. It ensures that stakeholders have access to understandable data and are informed of the rationale, assumptions, and implications behind key decisions. Transparency builds credibility and trust among affected communities.
Accountability	Involves establishing clear responsibilities and mechanisms to hold decision-makers answerable for their actions. This includes enabling public oversight, allowing for independent review, and ensuring that outcomes align with declared objectives. It empowers communities to monitor progress and ensure fair implementation.
Neutrality	Denotes impartiality in how procedures are conducted and decisions are made. It requires that no group receives preferential treatment and that processes are free from bias, political interference, or undue influence. Neutrality supports equitable treatment of all stakeholders.
Correctability	Refers to the ability of a system or process to accommodate revisions or appeals. It ensures that unjust or flawed decisions can be challenged and amended, thereby enhancing procedural resilience and legitimacy. Mechanisms for feedback, complaints, and appeals are central here.
Ethicality	Signifies alignment of processes with moral and ethical standards, including respect for human rights, cultural values, and environmental responsibilities. Ethicality ensures that the methods used are not only legal but also just and socially responsible.
Trustworthiness	Involves consistency, reliability, and adherence to stated principles over time. When institutions and actors demonstrate procedural integrity, they foster stakeholder confidence and long-term legitimacy of energy planning and implementation processes.
Voice	Concerns the inclusion of diverse perspectives and the opportunity for all affected groups—especially marginalized populations—to express their concerns, needs, and aspirations. It values deliberative dialogue and active listening as part of participatory justice.
Decision Control	Goes beyond mere consultation to ensure that stakeholders can exert real influence over decisions. This principle highlights the importance of shared authority, co-creation, and participatory governance structures where community input materially affects outcomes.
Capabilities	Refers to the need for empowering individuals and communities through adequate information, training, resources, and institutional support so they can effectively participate. This ensures that involvement is meaningful rather than symbolic.
Respect	Entails treating all participants with dignity, regardless of their background or viewpoint. It acknowledges different forms of knowledge, cultural practices, and lived experiences as legitimate and valuable to the decision-making process.
Politeness	Reflects courteous and professional interaction among participants. It promotes an inclusive and safe environment for dialogue, particularly important in emotionally charged or contested energy debates. Politeness reinforces mutual respect and constructive engagement.

2.2.4 *Restorative justice*

Restorative justice, as an emerging dimension within the energy justice framework, focuses on addressing past and present harms caused by energy-related decisions and processes, with the aim of restoring equity, trust, and social cohesion. While distributional, procedural, and recognition justice concentrate on fair access, participation, and representation, restorative justice takes a step further—it seeks to repair damage already inflicted on individuals, communities, and ecosystems as a result of unjust energy practices (Heffron and McCauley, 2017; Basil and Heffron, 2025).

At its core, restorative justice acknowledges that historical injustices—such as the displacement of local communities for energy infrastructure projects, environmental degradation caused by extractive industries, or the long-term exclusion of marginalized populations from energy governance—cannot be resolved merely by equitable forward-looking policies (Lewis et al., 2020). Instead, it calls for mechanisms that directly address these legacies. This may involve for example, formal apologies, financial or material compensation, land restoration, or renewed rights to participate in and benefit from the energy transition.

Importantly, restorative justice is rooted in dialogue and reconciliation. It emphasizes the need for open, inclusive, and respectful engagement between energy actors—governments, companies, and institutions—and the affected communities. Through such dialogue, harms are acknowledged, accountability is accepted, and pathways toward reparation are jointly identified. This approach is particularly relevant in contexts where communities have experienced systematic harm, such as indigenous populations whose territories have been disrupted by extractive energy projects or former industrial regions left economically vulnerable by sudden decarbonization efforts.

Restorative justice also reinforces the ethical foundation of energy transitions by placing emphasis on healing and rebuilding trust. It highlights the importance of rebuilding relationships between institutions and citizens—especially those who have experienced environmental or social injustices for decades. In doing so, it supports a more inclusive and morally responsible transition to sustainable energy systems.

A study by Hazrati and Heffron (2021), which investigated the role of restorative justice through a comparative analysis of fossil-fuel sectors in Iran, Malaysia, and Canada, demonstrates the potential of this justice dimension in strengthening the practical enforcement of energy justice. Their findings underscore the relevance of various legal instruments—such as Environmental Impact Assessment (EIA) procedures and the Energy Financial Reserve Obligation—that are grounded in restorative justice principles. As noted by the authors, the increasing implementation of such mechanisms indicates a tangible shift of restorative justice from theoretical discourse towards practical application in the energy sector.

However, additional research is required to advance the practical implementation of restorative justice, particularly in addressing the enduring consequences of environmental pollution, ecological degradation, and land dispossession. As it was noticed by Basil and Heffron (2025), this raises critical considerations surrounding accountability as a central element of restorative processes—specifically, the question of who should bear responsibility for the infringement of rights associated with energy development.

2.2.5 *Cosmopolitan Justice*

Cosmopolitan justice extends the normative framework of energy justice beyond national boundaries, emphasizing the ethical responsibility of addressing global inequalities in access to energy resources, technologies, and decision-making power. It is rooted in the principle that all individuals, regardless of citizenship or geographical location, possess equal moral worth and are entitled to fair behavior towards global energy systems. This perspective challenges the conventional nation-state focus of energy governance and calls for an ethical reorientation that recognizes shared obligations toward present and future populations across the globe (Maltais, 2008; Harris and Symons, 2010; Sovacool and Dworkin, 2015; Heffron et al., 2015).

In the context of energy transitions, cosmopolitan justice highlights the asymmetries between industrialized and developing countries in terms of historical emissions, technological capabilities, and financial resources. It draws attention to the disproportionate burdens carried by low-income countries and climate-vulnerable regions, which often contribute the least to global environmental degradation but are most affected by its consequences. From this viewpoint, global energy justice must involve mechanisms for international cooperation, technology transfer, and climate finance to support equitable energy development pathways.

Furthermore, cosmopolitan justice questions the legitimacy of energy policies and climate strategies that reinforce global power imbalances or neglect the voices of populations in the Global South. It emphasizes inclusive multilateralism, where all countries, particularly those historically marginalized, have a meaningful role in shaping global energy futures. This includes not only participation in high-level negotiations but also the recognition of diverse development models, knowledge systems, and cultural priorities.

The cosmopolitan lens also engages with intergenerational justice on a planetary scale. It stresses that the ethical obligation to prevent long-term ecological harm must not be confined to national interests but must be framed as a collective responsibility to humanity as a whole. In this sense, cosmopolitan justice acts as a moral compass for just energy transitions, encouraging countries, corporations, and international institutions to adopt policies that are environmentally sustainable, socially inclusive, and globally equitable.

Cosmopolitan justice adds a critical transboundary dimension to the energy justice discourse, as it urges recognition of shared vulnerability and interdependence, advocating for systemic reforms that ensure fair energy outcomes across both space and time (Heffron, 2020). As highlighted by Sokołowski and Heffron (2022), the responsibilities associated with energy governance extend beyond national boundaries, necessitating the integration of international dialogue and the involvement of multilateral institutions to mediate and resolve cross-border disputes. Accordingly, national energy strategies—whether in the form of policies, plans, programs, or projects—must account for their transboundary implications. These impacts may manifest at the regional level, for example, when neighboring countries are affected by infrastructure developments such as nuclear power plants (Sovacool et al., 2019), large-scale hydropower projects (Jayapalan and Ganesh, 2019), or intensive gas flaring (Aigbe et al., 2023). They may also arise from, for example, resource exploitation, including the extraction of rare earth minerals and associated labor conditions (Sovacool et al., 2019). At a global scale, such concerns are exemplified by the transboundary effects of greenhouse-gas (GHG) emissions. The Paris Agreement serves as a prominent example of cosmopolitan justice, illustrating collective responsibility in addressing shared environmental challenges.

Despite critiques such as those put forward by Ayllón (2025), who argues that the concept of cosmopolitan justice should be excluded from the energy justice framework due to its inherently anthropocentric orientation, the principle continues to feature in contemporary debates on global energy ethics. As energy systems become increasingly globalized, the relevance of the cosmopolitan dimension will only grow, reminding policymakers and scholars alike that justice must extend beyond borders. However, as it was highlighted by Wood et al. (2024), advancing the energy justice agenda requires deeper ethical engagement with the legitimacy of international inequalities. This entails critically examining whether disparities between countries can be morally justified and considering the implications of such assessments for shaping equitable energy policies and frameworks for global redistribution.

2.2.6 *Comparative Overview of Energy Justice Dimensions*

Table 2.2 summarizes the five core dimensions of energy justice—distributional, recognition, procedural, restorative, and cosmopolitan—into a concise comparative overview. For each dimension, it identifies the principal focus, outlines its defining principles, explicitly indicates spatial and temporal scopes to highlight the scale and timeframe over which each dimension primarily operates, and provides a representative example from energy policy or practice. This format enables a clear, side-by-side comparison of how different justice concerns manifest within energy systems.

While the majority of energy justice frameworks have traditionally centered on the triad of distributive, recognition, and procedural justice (e.g., Jenkins et al., 2016; Zhong et al., 2024; Bindi et al., 2025), recent scientific literature have increasingly highlighted the relevance of restorative and cosmopolitan justice as equally important components in advancing a more comprehensive and inclusive energy justice agenda.

Basil and Heffron (2025) investigated the intersection between energy justice and human rights frameworks, focusing on how these normative principles have

Table 2.2 Comparative overview of key energy justice dimensions. Source: developed by the author.

	Core focus	*Key principles*	*Temporal focus*	*Spatial focus*	*Examples in an energy context*
Distributional Justice	Fair distribution of energy-related benefits and burdens across society	• Equity in access to energy • Compensation for disproportionate impacts • Intergenerational fairness	Present and future	Local, regional, national	Unequal exposure to pollution from energy infrastructure, fuel poverty, and regional disparities in energy access
Recognition Justice	Acknowledging and respecting diverse identities, experiences, and rights	• Respect for cultural diversityo Elimination of misrecognition and discrimination • Valuing local knowledge	Historical and present	Local, regional	Inclusion of Indigenous worldviews in project planning; eliminating stereotypes in community energy programs
Procedural Justice	Inclusive, transparent, and fair decision-making processes	• Meaningful participationo Access to information • Accountability in governance	Present	Local, regional	Community engagement in renewable energy siting; transparent public consultations; inclusive energy policymaking
Restorative Justice	Addressing past harms and promoting repair, reconciliation, and social healing	• Reparation and compensationo Acknowledgment of injustice • Institutional accountability	Historical, present, future	Local, national	Restitution for displaced communities; restoration of damaged ecosystems; truth-telling about energy-related harms
Cosmopolitan Justice	Promoting global equity and transnational responsibility in energy governance	• Global fairness • Shared responsibility • Support for vulnerable nations and populations	Present and future	Global, transboundary	Technology transfer to the Global South; climate finance; equitable global decarbonization strategies

been operationalized within the energy sector to address pressing challenges such as energy poverty, environmental degradation, violations of community rights, and deficiencies in institutional accountability. Their findings reveal that while distributive and procedural dimensions of justice have gained increasing recognition, notable gaps remain—particularly regarding the integration of cosmopolitan and restorative justice perspectives into energy governance.

Cosmopolitan justice, which centers on the ethical consideration of transboundary impacts and global responsibilities, remains underdeveloped despite the inherently international nature of energy production, trade, and investment. As cross-border infrastructure projects and the activities of transnational energy corporations become more prevalent, the necessity for mechanisms that support international dialogue and accountability becomes more urgent. A failure to institutionalize cosmopolitan justice may leave systemic injustices unaddressed, especially those arising from extractive activities or infrastructure developments that transcend national borders yet disproportionately affect local populations.

Similarly, restorative justice—rooted in principles of repair and reconciliation—has not been sufficiently incorporated into energy justice debates. Originating from criminal justice theory, the restorative approach emphasizes addressing harm experienced by communities through inclusive, participatory processes that seek to restore social and environmental well-being. Although recent academic contributions have begun to explore its relevance in contexts of historical and structural energy-related injustices, particularly among marginalized groups, this remains an emerging strand within the broader energy justice literature (Basil and Heffron, 2025). The underutilization of restorative justice principles underscores a broader need to expand normative frameworks in energy policy to more fully account for historical harms and to promote inclusive, long-term resolution pathways.

Beyond these foundational and emerging principles, the academic literature continues to propose alternative conceptualizations of justice within the energy domain. For instance, Bal et al. (2023) introduce interactional justice as an additional dimension to the traditional framework. This principle focuses on the fairness of interpersonal treatment and the communicative quality of interactions between stakeholders and decision-makers. As such, it complements procedural justice by underscoring the importance of respectful, transparent, and empathetic engagement, particularly in the context of participatory processes central to just energy transitions.

Sovacool et al. (2016) proposed an alternative framework for energy justice that elaborates the core dimensions—distributive, procedural, and recognition justice—through a set of practical principles. For the distributive justice dimension, four interrelated principles can be assigned: availability (ensuring universal access to energy services), affordability (guaranteeing fair energy costs), intragenerational equity (fairness among individuals and social groups within the current generation), and sustainability, which reflects intergenerational equity by emphasizing the long-term environmental and social viability of energy systems. The dimension of procedural justice is articulated through the principles of due process, as a foundation for inclusive and equitable decision-making, and transparency and accountability, which stress the importance of clear communication and institutional responsibility in shaping energy policies and interventions. Finally, the principle of responsibility is positioned at the intersection of procedural and recognition justice, highlighting both the ethical duty to acknowledge and address the specific needs of marginalized groups and the institutional obligation to correct historic or systemic injustices.

2.3 Main Dimensions of Social Equity

Social equity in the energy sector refers to the fair and impartial distribution of the benefits and burdens arising from energy production, distribution, and consumption and constitutes a fundamental pillar for the sustainable development of the energy sector. The multidimensional nature of social equity has been increasingly recognized in energy policy and academic discourse, converging around four principal dimensions: access to modern energy services, affordability of energy costs, quality and reliability of supply, and participation in energy decision-making processes together with the recognition of the differentiated needs and circumstances of social groups, including low-income households, rural populations, women, the elderly, and minority communities. While social equity concerns are frequently examined within the framework of energy justice—particularly through its core dimensions of distributive, procedural, and recognition justice—this monograph distinguishes social equity according to specific dimensions: access, affordability, reliability, and participation. This approach aims to clarify the conceptual boundaries between social equity and energy justice, emphasizing the particular aspects covered by social equity in the context of the energy sector. Together, these dimensions offer a comprehensive lens through which the equity of energy systems can be evaluated and improved.

In many regions, structural inequalities in access to affordable and clean energy persist. Despite significant global efforts to expand electricity access over recent decades—resulting in approximately 925 million people gaining access since 2000, and nearly 40 countries achieving near-universal electrification—about 750 million individuals still lack access to electricity. Moreover, progress has been uneven, with only around 20% of those newly connected during this period residing in Africa, underscoring persistent regional disparities (International Energy Agency, 2024). Even in high-income countries, energy affordability remains a pressing issue. Eurostat (2025) data indicate that up to 10% of EU households were affected by energy poverty in 2023, disproportionately impacting women, elderly persons, and single-parent households. These disparities illustrate how energy systems can perpetuate broader socio-economic inequalities if social equity is not explicitly addressed in policy design and implementation.

2.3.1 Access

Access constitutes a fundamental dimension of social equity in the energy sector, representing the initial and essential condition for benefiting from modern energy services. While the physical connection to energy infrastructure is a necessary prerequisite, a nuanced understanding recognizes that access also involves geographical, technological, temporal, social, and institutional layers. Geographically, disparities persist between urban centers and rural or remote areas, where extending centralized grids is often economically and technically challenging. This spatial inequality underscores the importance of diversified energy delivery models, such as decentralized renewable energy systems, microgrids, and standalone solar home systems, which offer flexible solutions tailored to the specific needs and contexts of underserved communities. Technological access encompasses not only the physical availability of energy infrastructure but also the capacity to utilize modern, efficient, and clean energy technologies. A formal connection to the electricity grid does not inherently translate into the meaningful use of energy that enhances well-being or supports the goals of sustainable development. In many low-income and rural areas, the absence of affordable and appropriate end-use technologies—such as clean cooking solutions, efficient space heating systems, or electric appliances—continues to constrain energy access. This technological gap reinforces persistent health risks from indoor air pollution, accelerates local environmental degradation through unsustainable biomass harvesting, and sustains gendered inequalities in labor and time use (Ocen et al., 2024; Dirisu et al., 2024).

The World Health Organization (2025) reports that in 2023 approximately 2.1 billion individuals globally continue to lack access to clean and modern cooking technologies. Instead, these populations depend on the combustion of traditional solid fuels—such as firewood, charcoal, coal, animal dung, and agricultural residues—typically in inefficient cooking arrangements such as open fires. The use of such fuels in poorly ventilated indoor environments results in the emission of hazardous air pollutants, including fine particulate matter, carbon monoxide, and various toxic organic compounds. This form of household air pollution constitutes a severe environmental health risk, with particularly adverse effects for women who are most frequently exposed due to traditional gender roles in many low-income and rural areas. The impacts extend beyond public health, contributing significantly to environmental degradation, forest loss, and climate change through the release of greenhouse-gas (GHG) emissions and short-lived climate pollutants. Furthermore, the time and labor required to collect biomass fuels can reinforce cycles of poverty by limiting opportunities for education and income-generating activities.

The combustion of solid fuels for household purposes remains a dominant source of black carbon emissions globally, accounting for as much as 58% of the total, particularly in developing countries (World Bank, 2019). Among the various short-lived climate pollutants, black carbon represents the most substantial contribution from the residential energy sector. Cooking with traditional biomass or coal-based fuels alone is estimated to contribute approximately one-quarter of the total global emissions of black carbon that can be mitigated through targeted interventions. In certain regions of sub-Saharan Africa and South Asia, this share may rise to 60–80%, highlighting substantial regional disparities in emission profiles. From a mitigation perspective, addressing emissions of short-lived climate pollutants, particularly black carbon, presents an opportunity for achieving immediate climate-related and public health benefits. Unlike carbon dioxide, which remains in the atmosphere for extended periods and thus requires long-term strategies for stabilization, the atmospheric lifetime of black carbon is relatively short. Therefore, its reduction offers a rapid pathway to slow the rate of global warming and alleviate associated environmental risks (United States Environmental Protection Agency, 2025).

Beyond its contribution to atmospheric emissions, health risk, and climate change, the use of biomass for household energy also exerts significant pressure on natural ecosystems. The collection of wood for cooking and heating purposes is a major driver of forest degradation and land-use transformation, frequently leading to the fragmentation and loss of critical wildlife habitats. It is estimated that about one third of global wood fuel is extracted at rates that exceed natural regeneration, thereby classifying this portion of harvesting as environmentally unsustainable. Moreover, more than 275 million individuals reside in geographic regions identified as high-risk zones—often referred to as wood fuel "hotspots"—where over half of the wood harvested for energy purposes is obtained through practices that undermine ecological stability. These patterns of unsustainable biomass exploitation not only threaten biodiversity but also reduce the long-term resilience of local communities that depend on forest resources for their livelihoods (United States Environmental Protection Agency, 2025).

The temporal dimension of access is also equally significant: energy must be continuously and reliably available. Intermittent supply or frequent outages disproportionately affect low-income households who often lack alternatives, deepening their vulnerability. In addition, social and cultural barriers—including discriminatory practices, gender roles, and institutional biases—may restrict equitable access even where infrastructure exists (Johnson et al., 2020; Kelkar et al., 2121). Institutional frameworks and regulatory policies that enshrine energy as a human right and promote nondiscriminatory practices are crucial for addressing such barriers. Furthermore, legal and governance mechanisms that facilitate consumer protections, dispute resolution, and community-based management strengthen the substantive realization of access.

2.3.2 *Affordability*

Affordability is intrinsically linked to the broader notion of access and constitutes a pivotal element of social equity within the energy sector. It encompasses the capacity of individuals and households to secure and sustain necessary energy services—such as heating, lighting, cooking, and appliance usage—without undermining their ability to meet other essential needs, including food, housing, healthcare, and education. The interdependencies among energy access dimensions—namely availability, accessibility, affordability, and reliability—are empirically substantiated in the study by Mustapha et al. (2025). Their findings highlight that affordability serves as a critical determinant influencing other aspects of energy access; for instance, when energy services are financially out of reach, households may be unable to utilize existing infrastructure (accessibility), experience reduced consumption below subsistence levels (availability), or rely on unsafe or unreliable alternatives, thereby compounding vulnerabilities related to reliability.

In this context, affordability is not merely a function of energy pricing but is deeply embedded within wider socioeconomic structures and patterns of inequality. Direct energy-related expenditures—such as electricity and gas tariffs, taxes, fixed connection charges, and fees for metering or maintenance—can impose a disproportionate burden on low-income and vulnerable populations, who often spend a significantly higher share of their income on energy. This burden contributes to persistent energy poverty, a condition that exacerbates broader social exclusion and intensifies other social challenges, particularly in regions where income inequality is pronounced or where social support systems are underdeveloped.

However, the affordability extends beyond these direct expenses. Indirect costs are frequently overlooked in policy discourse yet play a substantial role in determining actual energy-related financial stress. These indirect costs may arise from the need to compensate for unreliable supply—such as through the purchase of backup generators, battery storage systems, or expensive fuels—or from structural inefficiencies related to poor housing quality and inadequate insulation, which increase energy consumption due to the low energy efficiency of the buildings. Furthermore, the use of polluting energy sources such as coal, wood, or kerosene in the absence of affordable alternatives contributes to health risks that turn into additional household expenditures, particularly for vulnerable groups such as children, the elderly, and those with preexisting conditions. These multidimensional burdens call for a holistic approach to affordability, one that integrates economic, infrastructural, and health considerations.

Affordability is also a dynamic and context-sensitive concept. It is affected by various temporal and structural factors, including seasonal fluctuations in energy demand—such as heating needs during winter—changes in household income levels due to employment shifts or macroeconomic instability (Che et al., 2021), and broader price volatility in global energy markets (Siksnelyte-Butkiene, 2022). For example, sudden spikes in energy prices driven by pandemics (Siksnelyte-Butkiene, 2021), geopolitical disruptions or supply chain constraints can disproportionately affect those with the least capacity to absorb shocks, thereby amplifying vulnerability (European Commission, 2022). A particularly illustrative case of the nexus between geopolitical instability and energy insecurity is the Russian invasion of Ukraine, which resulted in substantial volatility in global energy markets (Celi et al., 2022; Guarascio et al., 2025). The ensuing constraints on liquefied natural gas transportation were closely associated with short-term increases in energy deprivation and heightened levels of social tension. Empirical findings suggest that lower-income countries were disproportionately affected, exhibiting heightened vulnerability to fluctuations in gas supply and a greater likelihood of experiencing severe socio-political disruptions as a consequence (Zhang and Wang, 2024). Thus, policy interventions must be designed to be responsive and adaptive, ensuring that support mechanisms are sufficiently robust to withstand fluctuations and targeted enough to reach those most need it.

2.3.3 Reliability

Reliability, although often overlooked in social equity discourse within the energy sector, represents a critical dimension of equitable energy access. It entails both the continuous availability and the technical quality of energy services, which are fundamental to ensuring the well-being and daily functioning of consumers. For vulnerable and marginalized groups, reliable energy provision is essential not only as a convenience but as a prerequisite for meeting basic needs such as lighting, heating, refrigeration, medical device operation, and income-generating activities.

Interruptions in energy supply—manifesting as frequent outages, voltage fluctuations, or intermittent access—disproportionately affect households lacking the financial or technical capacity to mitigate these disruptions, such as those unable to afford backup generators, battery storage, or cleaner fuel alternatives. The implications of unreliable supply extend beyond inconvenience, posing tangible health risks in contexts where energy supports critical medical equipment or hygienic practices. For example, Messersmith et al. (2022) highlighted the substantial challenge posed by poor power quality to neonatal care in Special Neonatal Care Units in Central India, where equipment functionality is vital to infant survival.

Economic activities at the household level are similarly vulnerable. Small-scale enterprises and home-based work—often essential sources of income for low-income populations—are frequently disrupted by unreliable electricity. Additionally, technical quality parameters, including voltage stability and frequency regulation, play a crucial role in safeguarding household appliances and infrastructure. Deficiencies in these parameters contribute to premature appliance failure, resulting in increased replacement costs and exacerbating economic vulnerability.

Empirical evidence substantiates the broad consequences of unreliable electricity for households and businesses. For instance, Nduhuura et al. (2021) employed correlation and regression analyses to examine household-level factors associated with reported impacts of power outages. Their study identified that outages adversely affect household safety and security, food access, and availability of social services, in addition to causing damage to electrical appliances. Key factors significantly correlated with outage-related impacts included respondents' annual income, employment status, frequency of outages experienced, and household size. Moreover, socioeconomic disadvantage, high outage exposure, and larger household size were significant predictors of reporting such impacts. These findings underscore the importance of enhancing electricity reliability to improve household safety, reduce food loss, and contribute to broader food security objectives in developing regions.

Complementing these findings, empirical analysis by Cole et al. (2018), using firm-level data from 14 countries in the World Bank Enterprise Surveys, demonstrates a clear negative association between unreliable electricity supply and business sales. The effect was particularly pronounced among firms lacking backup generators. Their estimates suggest that aligning outage levels across Sub-Saharan Africa with those observed in South Africa could potentially increase overall firm sales by 85.1%, with an even greater increase of 117.4% for businesses without generator access. This evidence highlights the critical role of reliable electricity in supporting economic productivity and resilience among enterprises.

Furthermore, empirical evidence underscores that unreliable electricity supply—manifested through frequent power outages—can have significant environmental and health consequences, particularly in low-income settings. In response to electricity disruptions, households often resort to traditional biomass fuels such as firewood, charcoal, or kerosene as alternative energy sources. This adaptive behavior, while necessary for basic subsistence, contributes to increased indoor air pollution, posing severe health risks due to prolonged exposure to particulate matter and toxic emissions. Moreover, the reliance on wood-based fuels intensifies deforestation pressures and accelerates land degradation, thereby undermining local environmental sustainability. At a broader scale, the combustion of carbon-intensive traditional fuels during outages contributes to greenhouse-gas emissions, reinforcing global warming trends. These interconnected outcomes reflect the compounded risks of unreliable energy access, as observed in the Ethiopian context and more broadly in many low-income regions (Meles, 2020).

Reliability further encompasses the resilience of the energy system to withstand and recover from disturbances caused by natural disasters, technical malfunctions, fuel supply interruptions, or socio-political crises. As marginalized groups tend to be disproportionately exposed to these risks and are less equipped to adapt due to financial constraints or lack of access to backup solutions, social equity considerations necessitate targeted investments in robust and modernized infrastructure that prioritizes underserved regions and rural communities. In this case decentralized energy solutions, including microgrids and distributed generation, show potential in enhancing local reliability and reducing dependence on vulnerable centralized grids.

2.3.4 Participation

Participation constitutes a core procedural dimension of social equity, enabling diverse social groups—particularly those historically marginalized—to actively engage in shaping energy sector development. Moving beyond traditional top-down governance, socially equitable energy systems are founded on inclusive decision-making processes that provide all stakeholders with meaningful opportunities to voice their needs, concerns, and preferences, thereby influencing policies and practices that impact their lives and communities.

This focus on procedural fairness within social equity frameworks acknowledges that equity cannot be fully understood through distributive outcomes alone. Procedural fairness pertains to the transparency, inclusiveness, and legitimacy of decision-making mechanisms, emphasizing the meaningful involvement of all stakeholders, especially vulnerable and marginalized populations. Empirical evidence demonstrates that neglecting procedural dimensions often leads to perceptions of injustice, even when resource allocation appears equitable. Excluding affected communities from decision-making risks fosters distrust, reducing legitimacy, and undermining the sustainability of energy interventions. Therefore, procedural fairness is indispensable for a holistic conception of social equity in the energy sector, ensuring that just and sustainable energy systems are grounded not only in equitable resource distribution but also in fair and participatory governance processes.

In the context of social equity within the energy sector, procedural fairness encompasses several critical components that collectively ensure just and legitimate decision-making processes (Jenkins et al., 2016; Sovacool et al., 2019; Mueller and Keil, 2020; Shejale et al., 2025):

- *Participation should be active and inclusive,* ensuring that all stakeholders—particularly marginalized and vulnerable groups, who are often disproportionately affected by energy policies and projects—are meaningfully engaged throughout the entire decision-making process. This means not only providing opportunities for input but also enabling meaningful influence over outcomes. Inclusive participation addresses power imbalances by ensuring that voices traditionally excluded—such as low-income households, indigenous populations, and women—are integrated into policy formulation, planning, and implementation phases. Such engagement helps to identify diverse needs, contextual challenges, and locally relevant solutions, thereby improving the equity and effectiveness of energy interventions.
- Decision-makers should ensure *transparency* through the clear, consistent, and accessible disclosure of decision-making criteria, procedures, and underlying rationales to all stakeholders. Transparency enhances trust by reducing information asymmetries and enabling stakeholders to comprehend how and why decisions are made. It also allows civil society and affected communities to hold authorities accountable. Transparent processes minimize opportunities for corruption or undue influence and support evidence-based policymaking aligned with equitable outcomes.
- Decision-makers should uphold *impartiality* by ensuring that decisions are free from bias, favoritism, or discriminatory practices that may privilege certain groups over others. Impartiality ensures that decisions are guided by objective criteria and ethical standards rather than political, economic, or social pressures that could perpetuate inequities. Maintaining impartiality also strengthens the legitimacy of energy governance institutions by fostering perceptions of neutrality and fairness.

- All rules, procedures, and standards should be applied *consistently* and uniformly across all affected parties. Consistency prevents arbitrary or ad hoc decision-making that could lead to unequal treatment or unpredictability. In the energy sector, consistent application of regulations related to access, tariffs, and service quality is essential to guarantee that no groups are disproportionately disadvantaged or excluded. Consistency also facilitates long-term planning and investment by providing stakeholders with clear expectations about the regulatory environment.
- Decision-makers and institutions should be held *accountable* for their actions and outcomes, with mechanisms in place to provide redress and correction in cases of unfair or erroneous decisions. Accountability frameworks may include, for example, formal complaint procedures or independent regulatory bodies. In energy governance, accountability protects consumer rights, promotes ethical behavior, and reinforces trust in institutions. It is especially critical where vulnerable populations depend on energy services for fundamental needs such as health, education, and economic livelihoods.

Collectively, these elements of procedural fairness form the foundation for social equity in energy systems by ensuring that the processes guiding energy distribution and policymaking are just, inclusive, and transparent. Addressing procedural fairness is therefore indispensable for achieving not only equitable outcomes but also legitimacy and sustainability in energy transitions.

Consequently, participation should entail more than formal consultation or symbolic representation. It requires structured and sustained stakeholders engagement, wherein individuals and communities are empowered to contribute knowledge, challenge prevailing assumptions, and co-shape priorities in areas such as infrastructure planning, service delivery models, or location of new energy facilities. Such involvement helps ensure that energy development reflects the expectations, cultural values, and socioeconomic realities of diverse groups of population, thereby reducing the likelihood of inequitable outcomes or social conflict.

However, the ability to participate effectively is often constrained by multiple overlapping barriers. These include informational asymmetries—where technical knowledge is concentrated among a few people or an exact group of people; limited access to education, knowledge, or digital technologies; linguistic and cultural divides that hinder communication; or entrenched power hierarchies that marginalize certain voices while amplifying others. In many contexts, procedural exclusion is not merely incidental but structurally embedded, reinforcing social inequities further and undermining the democratic legitimacy of energy governance and application of social equity principles in shaping the energy sector. Empirical research, such as that conducted by Licht and Håkansson (2025), emphasizes the challenges in achieving genuine procedural fairness—even in favorable contexts—and underscores the necessity of governance frameworks that effectively address power imbalances, mediate conflicting interests, and promote equitable and inclusive decision-making processes.

In this regard, participation is directly linked to the acknowledgment of diverse social identities, experiences and variety of needs. Recognizing of various marginalized groups—such as low-income households, ethnic minorities, indigenous communities, and women—not only enhances procedural justice but also affirms their right to shape their pathways in accordance with their specific needs and wishes. The inclusion of such perspectives contributes to a pluralistic governance environment, capable of generating more socially responsive and context-sensitive energy solutions.

Moreover, assurance of participation also holds broader societal value beyond the energy sector. It can strengthen social cohesion, build trust in governmental institutions, and enhance knowledge level among communities. Through participatory engagement, communities often develop stronger networks, local leadership capacities, and shared ownership over energy initiatives—outcomes that enhance resilience and long-term sustainability.

Finally, although the concept of procedural justice is central to both social equity and energy justice discourses, its interpretation and scope vary across these two frameworks. Within the social equity paradigm, procedural justice primarily emphasizes equal and nondiscriminatory treatment, focusing on transparent, consistent, and inclusive administrative processes that ensure fair access to public services and participation opportunities. It tends to operate within existing institutional structures, aiming to improve representativeness and responsiveness, particularly for disadvantaged or underserved populations. In contrast, the energy justice framework conceptualizes procedural justice more normatively and structurally, seeking not only inclusion but also empowerment of historically marginalized groups in energy decision-making processes. Here, procedural justice includes deliberative participation, recognition of diverse knowledge systems, and agency in shaping energy transitions. It is concerned not just with how procedures are carried out, but also with transforming institutional power dynamics that influence whose voices are heard and whose interests are prioritized. Thus, while social equity focuses on improving procedural fairness within existing systems, energy justice seeks to reconfigure those systems to achieve more equitable and democratic energy governance.

2.3.5 Comparative Overview of Social Equity in the Energy Sector Dimensions

Table 2.3 presents a comparative summary of the key elements of social equity dimensions examined in this sub-chapter. It synthesizes the four central dimensions—access, affordability, reliability, and participation—by outlining their principal areas of focus, commonly employed indicators for evaluating equity-related outcomes, and selected examples illustrating their practical application in the energy sector. This structured overview allows to understand how these dimensions are conceptually defined and implemented in practice.

Table 2.3 Comparative overview of social equity dimensions in the energy sector. Source: developed by the author.

	Core focus	*Key principles*	*Typical indicators*	*Examples of applications*
Access	Physical and institutional availability of energy services	Universal service, nondiscrimination, territorial equity	• Population with electricity access • Urban–rural access gap • Connection lead times • Availability of energy infrastructure	• Expansion of off-grid solar home systems in remote areas • Electrification of informal settlements • National rural-electrification programs
Affordability	Energy cost burden on households	Fair pricing, social protection, targeted subsidies, and targeted energy efficiency measures	• Proportion of income allocated to energyo Disconnections for nonpayment • Average household energy bill • Thermal comfort satisfaction	• Social tariff schemes for low-income groupso Energy efficiency measures • Subsidies and targeted transfers
Reliability	Consistency and quality of energy supply	Service continuity, technical adequacy, resilience	• Frequency and duration of outageso Voltage stability • Consumer complaint rates	• Generators for critical infrastructureo Microgrids and decentralized energy systems • Energy storage integration to stabilize renewables
Participation	Fair engagement in energy-related decisions	Procedural justice, transparency, inclusiveness, empowerment	• Number of public consultations • Diversity in stakeholder participation • Existence of grievance mechanisms	• Community consultations and discussions on new energy infrastructure projects • Stakeholder engagement platforms • Energy literacy improvement workshops

2.3.6 Linking Social Equity Dimensions to Justice Frameworks

The four key dimensions of social equity in the energy sector—access, affordability, reliability, and participation—can also be understood within the well-established theoretical framework of energy justice—distributional, procedural, and recognitional justice. Specifically:

- *Distributional justice* corresponds to *access, affordability, and reliability*, emphasizing the fair allocation of energy resources and services to ensure that all individuals can obtain, afford, and depend on an adequate energy supply.
- *Procedural justice* aligns with *participation*, highlighting the importance of inclusive, transparent, and fair decision-making processes that empower affected communities and stakeholders to influence energy governance and policy.
- *Recognitional justice* pertains to *acknowledging marginalized groups' experiences*, ensuring that equity efforts go beyond equal treatment to recognize and address the unique social, cultural, and historical contexts of those disproportionately affected by energy poverty and exclusion.

Such an integrated perspective allows for a more comprehensive understanding of social equity in the energy sector by linking tangible, operational dimensions with underlying justice principles. This connection highlights that achieving equity goes beyond addressing material and service-based disparities—it also requires recognizing inclusive participation and the unique experiences of marginalized groups. By combining these aspects, it can be better understood the complex and interconnected nature of energy inequities.

2.4 Understanding the Nexus Between Energy Justice and Social Equity

The concepts of energy justice and social equity, though closely related, represent distinct analytical frameworks within the energy sector (Table 2.4). Energy justice is a normative and multidimensional approach that emphasizes fairness across the distribution of energy benefits and burdens, the inclusiveness of decision-making processes, and the recognition of historically marginalized stakeholders. It draws on theoretical foundations from environmental justice, political ecology, and human rights, positioning itself as a transformative agenda for addressing systemic and intergenerational injustices. In contrast, social equity in the energy sector focuses more narrowly on ensuring fair access to energy services, particularly through distributive mechanisms aimed at reducing disparities in affordability, infrastructure availability, and service reliability. While energy justice often operates at the level of structural reform and participatory governance—incorporating reparative policies and just transition principles—social equity is primarily operationalized through concrete

Table 2.4 Comparison of energy justice and social equity in the energy sector. Source: developed by the author.

	Energy justice	*Social equity*
Definition	A normative framework ensuring fairness in energy systems, encompassing distribution, procedures, and recognition of all stakeholders	Focuses on equitable access to energy services and affordability, aiming to reduce socioeconomic disparities
Theoretical foundations	Rooted in environmental justice, human rights, and political ecology, emphasizing procedural and recognitional justice	Based on welfare economics and social justice theories, concentrating on distributive justice and policy implementation
Core components	Distributive justice, procedural justice, recognitional justice, and recently restorative and cosmopolitan justice	Primarily concerned with distributive justice (access, affordability, service equality) and procedural fairness, acknowledging marginalized groups' experiences
Scope and application	Addresses systemic inequalities, including historical exclusion and participation in energy governance	Concentrates on current inequalities in energy access and cost burden at household and community levels
Policy instruments	Inclusive decision-making, stakeholder engagement, reparative policies, and just transition mechanisms	Social tariffs, targeted subsidies, energy efficiency programs, and infrastructure investment in underserved areas
Analytical approach	Multidisciplinary, combining normative ethics, law, and policy analysis	Empirical, quantitative, and policy evaluation focusing on measurable equity outcomes
Typical issues addressed	Fair siting of energy infrastructure, inclusive governance, and addressing intergenerational justice	Energy poverty reduction, affordability, and equitable access to modern energy services
Examples of practice	Community involvement in renewable energy projects, rights-based energy access, reparations for exclusion	Provision of social tariffs, rural-electrification projects, and subsidies for low-income households
Outcome orientation	Seeks transformative change in energy systems to achieve long-term sustainability and justice for all stakeholders	Aims to ensure equitable energy outcomes, reducing disparities and promoting social inclusion in energy consumption

policy tools such as social tariffs, targeted subsidies, and support programs for vulnerable populations. Both frameworks aim to enhance the fairness of energy systems, but they differ in scope, analytical orientation, and outcome targets. Understanding their complementarities is critical for designing inclusive and resilient energy policies, particularly in contexts where both structural injustices and immediate socioeconomic disparities persist.

To enhance more clarity in understanding the conceptual relationship between energy justice and social equity, these two frameworks can be also examined through the dimensions they encompass. Table 2.5 presents a comparative overview of their principal dimensions, highlighting both the connections and differences between them.

Table 2.5 Comparison and relationship between energy justice and social equity dimensions in the energy sector. Source: developed by the author.

Energy justice dimension	*Core focus*	*Related social equity dimension*	*Relationship / explanation*	*Key differences*
Distributional justice	Fair allocation of benefits and burdens from energy systems (costs, risks, access)	Access, Affordability, Reliability	Primarily relates to equitable access to energy services and affordability of energy. Reliability (consistent and quality energy supply) is included as an essential part of equitable distribution because inconsistent service imposes burdens.	Energy justice treats reliability as part of distributional fairness, while social equity highlights reliability as a standalone dimension due to its direct impact on quality of life.
Recognition justice	Acknowledgement and respect for diverse identities, cultures, and rights	Participation (overlaps with recognition of marginalized groups)	Ensures all groups, especially marginalized ones, are recognized in energy decisions, directly influencing participation and inclusivity.	Energy justice frames recognition broadly as respect and cultural inclusion, while social equity operationalizes this through participation in decision-making.
Procedural justice	Fairness in decision-making processes, transparency, and access to information	Participation	Overlaps in emphasizing inclusive, fair processes where stakeholders can engage and influence energy policies and projects.	Social equity focuses more on participation as a practical tool for inclusion, whereas energy justice frames procedural justice as a normative principle.
Restorative justice	Repairing harms caused by energy injustices, accountability, and reparation	Not typically included	Not a standard dimension in social equity, which tends to focus on current equitable outcomes rather than addressing past harms or reparations.	Restorative justice is normative and forward-looking in energy justice, addressing historic harms, which is generally outside the scope of social equity.
Cosmopolitan justice	Global justice considerations beyond national borders	Not included	Cosmopolitan justice addresses transboundary and global impacts, such as climate change and global energy trade, beyond local or national social equity concerns.	Social equity generally focuses on national and local scales, not the global or international dimension covered by cosmopolitan justice.

Building upon the distinct yet complementary nature of energy justice and social equity, their integration within energy policy and planning processes can substantially improve the inclusivity, legitimacy, and effectiveness of energy sector development. While energy justice provides a comprehensive normative framework aimed at addressing systemic and intergenerational injustices through principles of distributive, procedural, and recognition justice, social equity offers practical instruments that directly target disparities in access, affordability, and service reliability for marginalized and vulnerable groups.

Incorporating social equity into energy governance facilitates the identification of specific populations disproportionately affected by energy poverty or infrastructural deficits, enabling the design of targeted interventions such as social tariffs, subsidies, and improved service quality. Such a focus ensures that immediate inequities are addressed, thereby improving energy accessibility and social well-being at the community level.

Conversely, the broader framework of energy justice guides the pursuit of transformative reforms that reshape energy systems toward greater fairness, inclusiveness, and accountability. By emphasizing inclusive decision-making processes and the recognition of historically marginalized stakeholders, energy justice expands the policy lens beyond distributional concerns to encompass the political and ethical dimensions of energy sector development.

Together, these frameworks reinforce one another: social equity's focus on tangible outcomes and specific vulnerable groups complements energy justice's emphasis on structural change and participatory governance. This synergy enhances the capacity of energy policies to simultaneously address urgent socioeconomic disparities and embed long-term fairness and resilience in energy systems.

Empirical evidence suggests that integrating energy justice and social equity principles is particularly vital in contexts of rapid energy sector transformation (Chapman et al., 2016), where technological innovation, regulatory reform, and environmental sustainability must be harmonized with social inclusion and fairness. Therefore, adopting a dual approach not only facilitates more equitable access to energy services but also fosters broader societal acceptance and legitimacy of energy policies, ultimately contributing to sustainable and just energy sector development.

References

Aigbe, G.O.; Cotton, M.; Stringer, L.C. Global gas flaring and energy justice: An empirical ethics analysis of stakeholder perspectives. Energy Research & Social Science, 2023, 99, 103064.

Ayllón, L.M.S. Debates on the future of energy justice: Re-grounding the triumvirate. Environmental Science & Policy, 2025, 167, 104047.

Bal, M.; Stok, M.; Bombaerts, G.; Huijts, N.; Schneider, P.; Spahn, A.; Buskens, V. A fairway to fairness: Toward a richer conceptualization of fairness perceptions for just energy transitions. Energy Research & Social Science, 2023, 103, 103213.

Basil, C.V.; Heffron, R. A systematic review of the intersection between energy justice and human rights. Renewable and Sustainable Energy Reviews, 2025, 221, 115892.

Bindi, J.; Bartolomei, F.; Pellegrini-Masini, G.; Agostini, A.; Padovan, D. Assessing social impacts and Energy Justice along green hydrogen supply chains: a capability-based framework. Energy Research & Social Science, 2025, 126, 104149.

Campos, I.; Wittmayer, J.M.; Hielscher, S.; Oliveira, F.; Moreira Alves, F.; Progscha, S.; Wientjes, A. Just participation in wind energy: the role of social innovations. Renewable and Sustainable Energy Reviews, 2025, 209, 115146.

Celi, G., Guarascio, D., Reljic, J., Simonazzi, A., & Zezza, F. The asymmetric impact of war: Resilience, vulnerability and implications for EU policy. Intereconomics, 2022, 57(3), pp. 141–147.

Chapman A, McLellan B, Tezuka T. Strengthening the Energy Policy Making Process and Sustainability Outcomes in the OECD through Policy Design. Administrative Sciences, 2016; 6(3), p. 9. https://doi.org/10.3390/admsci6030009

Che, X.; Jiang, M. Economic Policy Uncertainty, Financial Expenditure and Energy Poverty: Evidence Based on a Panel Threshold Model. Sustainability, 2021, 13, 11594.

Cole, M.A.; Elliott, R.J.R.; Occhiali, G.; Strobl, E. Power outages and firm performance in Sub-Saharan Africa. Journal of Development Economics, 2018, 134, pp. 150–159.

Dirisu, J.O.; Oyedepo, S.O.; Olawole, O.C.; Somefun, T.E.; Peter, N.J.; Babatunde, D.; Nwaokocha, C.N.; Onokwai, A.O.; Obanor, E.; Alam, M.M.; Kale, S.A. A comprehensive review of biofuel utilization for household cooking in developing countries: Economic and environmental impacts. Process Safety and Environmental Protection, 2024, 191, pp. 585–604.

Ellis, G.; Schneider, N.; Wüstenhagen, R. Dynamics of social acceptance of renewable energy: an introduction to the concept. Energy Policy, 2023, 181, 113706.

European Commission. REPowerEU: Joint European Action for more affordable, secure and sustainable energy. Communication from the Commission to the European Parliament, the European Council, the Council, the European Economic and Social Committee and the Committee of the Regions. 2022, Strasbourg, 8.3.2022, COM(2022) 108 final.

Eurostat. Database. Income and living conditions. 2025. Available online at: https://ec.europa.eu/eurostat/databrowser/explore/all/popul?lang=en&subtheme=livcon.ilc&display=list&sort=category [Assessed on 09-05-2025].

Fraser, N. From Redistribution to Recognition? Dilemmas of Justice in a "Post-Socialist" Age. New Left Review, 1995, 212, pp. 68–93.

Guarascio, D., Reljic, J.; Zezza, F. Energy vulnerability and resilience in the EU: concepts, empirics and policy. J. Ind. Bus. Econ., 2025. https://doi.org/10.1007/s40812-025-00340-9

Harris, PG.; Symons, J. Justice in adaptation to climate change: cosmopolitan implications for international institutions. Environmental Politics, 2010, 19(4), pp. 617–636.

Hazrati, M.; Heffron, R.J. Conceptualising restorative justice in the energy Transition: Changing the perspectives of fossil fuels. Energy Research & Social Science, 2021, 78, 102115.

Heffron, R.J. Energy justice–the triumvirate of tenets revisited and revised. Journal of Energy & Natural Resources Law, 2023, 42(2), pp. 227–233.

Heffron, R.J. The role of justice in developing critical minerals. The Extractive Industries and Society, 2020, 7(3), pp. 855–863.

Heffron, R.J.; McCauley, D. The concept of energy justice across the disciplines. Energy Policy, 2017, 105, pp. 658–667.

Heffron, R.J.; McCauley, D.; Sovacool, B.K. Resolving society's energy trilemma through the Energy Justice Metric. Energy Policy, 2015, 87, pp. 168–176.

International Energy Agency. World Energy Outlook 2024. IEA Publications, 2024, 396 p. Available online at: https://iea.blob.core.windows.net/assets/140a0470-5b90-4922-a0e9-838b3ac6918c/WorldEnergyOutlook2024.pdf [Accessed on 16-05-2025].

Jayapalan, C.; Ganesh, L.S. Environmentalists and their conflicts with Energy Justice Concept of "Power-Environ" in the Athirappilly HEPP in Kerala. Energy Policy, 2019, 129, pp. 215–229.

Jenkins, K.; McCauley, D.; Heffron, R.; Stephan, H.; Rehner, R. Energy justice: A conceptual review. Energy Research & Social Science, 2016, 11, pp. 174–182.

Jenkins, K.E.H.; Sovacool, B.K.; Mouter, N.; Hacking, N.; Burns, M.K.; McCauley, D. The methodologies, geographies, and technologies of energy justice: a systematic and comprehensive review. Environmental Research Letters, 2021, 16, 043009.

Johnson, O.W.; Han, J.Y.C.; Knight, A.L.; Mortensen, S.; Aung, M.T.; Boyland, M.; Resurrección, B.P. Intersectionality and energy transitions: A review of gender, social equity and low-carbon energy. Energy Research & Social Science, 2020, 70, 101774.

Kelkar, G.; Nathan, D. Cultural and Economic Barriers in Switching to Clean Cooking Energy: Does Women's Agency Make a Difference? Energies, 2021, 14, 7242.

Kime, S.; Jacome, V.; Pellow, D.; Deshmukh, R. Evaluating equity and justice in low-carbon energy transitions. Applied Energy, 2023, 18(12), 123003.

Korsnes, M.; Throndsen, W. Smart energy prosumers in Norway: critical reflections on implications for participation and everyday life. Journal of Cleaner Production, 2021, 306, 127273.

Lewis, J.; Hernández, D.; Geronimus, AT. Energy efficiency as energy justice: addressing racial inequities through investments in people and places. Energy Efficiency, 2020, 13, pp. 419–432.

Licht, K.D.; Håkansson, M. Justice Under the Sun: Evaluating Procedural Justice in Large-Scale Solar Park Development. Sustainable Development, 2025, https://doi.org/10.1002/sd.3562

Maltais, A. Global warming and the cosmopolitan political conception of justice. Environmental Politics, 2008, 17(4), pp. 592–609.

Meles, T.H. Impact of power outages on households in developing countries: Evidence from Ethiopia. Energy Economics, 2020, 91, 104882.

Messersmith L.J, Ladha A, Kolhe C.; Patel, A.; Summers, J.S.; Rao, S.R.; Das, P.; Mohammady, M.; Conant, E.; Ramanathan, N.; Hibberd, P.L. Poor power quality is a major barrier to providing optimal care in special neonatal care units (SNCU) in Central India. Gates Open Research, 2022, 6, p. 58.

Mueller, C.E.; Keil, S.I. Measuring perceived procedural fairness in the context of power grid expansion A Rasch modeling approach. International Journal of Energy Sector Management, 2020, 14, 1, pp. 85–107.

Mustapha, R.A.; Devidas, A.R.; Sharma, A. Navigating the energy transition: insights from a systematic scoping review of challenges and pathways in developing nations. Energy Research & Social Science, 2025, 125, 104098.

Nduhuura, P.; Garschagen, M.; Zerga, A. Impacts of Electricity Outages in Urban Households in Developing Countries: A Case of Accra, Ghana. Energies, 2021, 14, 3676.

Ocen, S.; Nkurunziza, G.; Bagire, V.; Echegu, S.; Ssekakubo, J.; Atukunda, R. Unveiling factors influencing choice of clean cooking solutions among households: a systematic review of literature. Frontiers in Sustainability, 2024, 5, 1452900.

Oliveira, F.G.; Campos, I.; Alves, F.M. A quiet public? Procedural justice in Portuguese wind energy governance. Energy Research & Social Science, 2025, 126, 104147.

Pellegrini-Masini, G.; Pirni, A.; Maran, S. Energy justice revisited: A critical review on the philosophical and political origins of equality. Energy Research & Social Science, 2020, 59, 101310.

Ruano-Chamorro, C.; Gurney, G.G.; Cinner, J.E. Advancing procedural justice in conservation. Conservation Letters, 2022, 15(3), e12861.

Ryder, S.; Walker, C.; Batel, S.; Devine-Wright, H.; Devine-Wright, P.; Sherry-Brennan, F. Do the ends justify the means? Problematizing social acceptance and instrumentally-driven community engagement in proposed energy projects. Socio-Ecological Practice Research, 2023, 5 (2), pp. 189–204.

Shejale, S.; Zhan, M.X.; Sahakian, M.; Aleksieva, R.; Biresselioglu, M.E.; Bogdanova, V.; Cardone, B.; Epp, J.; Kirchler, B.; Kollmann, A.; Liste, L.; Massullo, C.; Schibel, K.L. Participation as a pathway to procedural justice: A review of energy initiatives across eight European countries. Energy Research & Social Science, 2025, 122, 103982.

Siksnelyte-Butkiene, I. Combating Energy Poverty in the Face of the COVID-19 Pandemic and the Global Economic Uncertainty. Energies 2022, 15, 3649.

Siksnelyte-Butkiene, I. Impact of the COVID-19 Pandemic to the Sustainability of the Energy Sector. Sustainability, 2021, 13, 12973.

Simcock, N. Procedural justice and the implementation of community wind energy projects: a case study from South Yorkshire, UK. Land Use Policy, 2016, 59, pp. 467–477.

Sokołowski, M.M.; Heffron, R.J. Defining and conceptualising energy policy failure: The when, where, why, and how. Energy Policy, 2022, 161, 112745.

Sovacool, B.; Heffron, R.J.; McCauley, D.; Goldthau, A. Energy decisions reframed as justice and ethical concerns. Nature Energy, 2016, 2016, 1, 16024.

Sovacool, B.K.; Dworkin, M.H. Energy justice: Conceptual insights and practical applications. Applied Energy, 2015, 142, pp. 435–444.

Sovacool, B.K.; Martiskainen, M.; Hook, A.; Baker, L. Decarbonization and its discontents: a critical energy justice perspective on four low-carbon transitions. Climate Change, 2019, 155(4), pp. 581–619.

Tarekegne, B. Just electrification: Imagining the justice dimensions of energy access and addressing energy poverty. Energy Research & Social Science, 2020, 70, 101639.

United States Environmental Protection Agency. Understanding the Issue: Household Energy and Clean Air. 2025. Available online at: https://www.epa.gov/indoor-air-quality-iaq/understanding-issue-household-energy-and-clean-air [Assessed on 02-06-2025].

Wood, N.; van Uffelen, N.; Frigo, G.; Melin, A.; Milchram, C.; Lee, J.; Bessa, S. Strengthening the foundations of energy justice scholarship: What can philosophy contribute? Energy Research & Social Science, 2024, 117, 103699.

World Bank. Clean Cooking: Why it Matters. Available online at: https://www.worldbank.org/en/news/feature/2019/11/04/why-clean-cooking-matters [Assessed on 02-06-2025].

World Health Organization. World health statistics 2025: Monitoring health for the SDGs, Sustainable Development Goals, 2025, p. 88. Available online at: https://iris.

who.int/bitstream/handle/10665/381418/9789240110496-eng.pdf?sequence=1 [Assessed on 02-06-2025].

Zhang, S.; Wang, L.H. The Russia-Ukraine war, energy poverty, and social conflict: An analysis based on global liquified natural gas maritime shipping. Applied Geography, 2024, 166, 103263.

Zhong, F.L.; Tian, J.W.; Zhao, C.X.; Zha, S.; Chen, X.; Zhang, Y.H. Assessing energy justice in climate change policies: an empirical examination of China's energy transition. Climate Policy, 2024, 24(3), pp. 362–377.

CHAPTER 3

Energy Transition and Social Justice

3.1 Understanding the Just Transition Framework

The concept of a just transition has emerged as a critical component of contemporary energy and climate policy, reflecting growing recognition that the shift to a low-carbon economy must be both environmentally sustainable and socially inclusive. While the energy transition refers to the structural transformation of energy systems away from fossil fuels toward renewable and cleaner energy sources, the just transition framework extends this scope by incorporating equity and justice considerations into the planning and implementation of low-carbon policies. In this context, the transition is not only a matter of technological change or environmental necessity but a broader socioeconomic transformation requiring deliberate attention to fairness, participation, and protection of vulnerable groups.

The just transition framework originates from the labor movement and was initially developed to ensure that workers in carbon-intensive sectors are not left behind as economies decarbonize. Over time, its application has broadened to encompass multiple dimensions of justice, including income equity, regional disparities, access to clean energy, and procedural inclusion. This multidimensionality positions the just transition as an integrative policy framework that aims to align climate objectives with social and economic development goals.

Justice-related challenges in low-carbon transitions emerge not solely from the contraction of fossil-fuel-based industries, but also from the unequal allocation of associated benefits. Evidence suggests that the growth of the renewable energy sector—often positioned as a key driver of "green job" creation—has yet to yield significant advantages for historically marginalized populations. For example, representation of women and ethnic minorities remains below average across many renewable energy occupations. As noted by Carley and Konisky (2020), the sector continues to exhibit a pronounced gender imbalance, particularly within high-skilled and well-compensated technical roles.

I. Siksnelyte-Butkiene, D. Streimikiene, *Social Equity in the Energy Sector*,
https://doi.org/10.1007/978-3-032-22224-4_3

Additionally, the uneven spatial deployment of low-carbon technologies and the localized externalities linked to energy system changes raise substantive concerns regarding environmental justice and the persistence of energy poverty.

Williams and Doyon (2019) emphasize that justice must be integral to sustainability transitions, asserting that a transition lacking fairness cannot be considered sustainable. Within this context, the notion of a just transition has gained prominence as a framework that centers on equity considerations in climate and energy policymaking, and it is increasingly acknowledged as a critical pillar of effective low-carbon governance.

As it was mentioned before, the notion of a just transition has its origins in the advocacy of labor movements and environmental justice groups dating back to the 1970s, particularly among trade unions calling for the protection of workers affected by environmental regulations. Nonetheless, it was not until the early 2000s that the concept began to receive more structured attention in academic research (Wang and Lo, 2021). Since then, interest in the topic has steadily increased, with contributions emerging from various disciplinary perspectives. This broad engagement has led to a range of interpretations and applications, contributing to significant variation in how the concept is understood. Despite its widespread use in both policy and academic discussions, a single, universally accepted definition or coherent analytical framework has yet to emerge (Henry et al., 2020). Consequently, the term is now used in diverse ways depending on context, which poses challenges for effective communication, policy alignment, and cross-contextual comparison (Wang and Lo, 2021).

When reviewing the main approaches to the concept of just transition, it becomes evident that numerous scholars highlight its origins in labor movements, which initially emerged in response to environmental regulations and called for social protection measures and equitable outcomes for affected workers and communities. As a result, just transition is frequently conceptualized as a process that inherently links the decarbonization of the energy system with broader economic restructuring, particularly in terms of employment, industrial change, and social equity.

Another strand of the literature conceptualizes just transition in close alignment with more established justice-related frameworks, particularly those of environmental, climate, and energy justice. The roots of environmental justice can be traced to long-standing social movements that have historically advocated for the rights and well-being of vulnerable communities disproportionately exposed to environmental harm. These movements have consistently argued that environmental burdens—such as pollution, ecological degradation, and exposure to hazardous sites—should not fall unevenly on marginalized populations, and that all individuals must have equal opportunities to participate in the formulation and implementation of environmental policy.

From this perspective, environmental justice expands beyond distributive concerns, incorporating both procedural and recognition dimensions of justice. It acknowledges that political, economic, and cultural inequalities are fundamental drivers of environmental injustice and seeks to identify not only what inequities exist, but also who is affected and in what ways. Accordingly, the concepts of recognition and participation have become integral components of environmental justice discourse. Recognition justice refers to the need for acknowledging diverse identities, experiences, and knowledge systems—particularly those of historically marginalized groups—while procedural justice stresses inclusive and transparent decision-making processes.

Scholars underscore that these three dimensions—distributive, recognition, and procedural justice—are not mutually exclusive but rather inherently interconnected (McCauley and Heffron, 2018). A fair distribution of environmental benefits and burdens cannot be achieved without first recognizing affected communities and ensuring their meaningful engagement in policy processes. As such, the integration of these justice dimensions is foundational to both the theoretical development and practical implementation of a just transition (Schlosberg, 2004; Wang and Lo, 2021).

The concept of climate justice has developed alongside that of environmental justice, expanding its focus to highlight how climate change exacerbates existing social inequalities. Climate justice advocates emphasize that addressing climate change requires not only environmental protection but also socially just outcomes. The same central elements are highlighted in the climate justice approach, which include compensatory justice—acknowledging and addressing historical harm; distributive justice—ensuring fair allocation of responsibilities and benefits in both present and future contexts; and procedural justice—promoting inclusive and transparent decision-making processes (Okereke, 2010).

At the core of energy justice lies the analytical focus on how the advantages and disadvantages associated with energy systems are allocated among different groups in society. Most conceptual models of energy justice rest upon four foundational dimensions: distributive justice, procedural justice, recognition justice, and restorative justice.

Consequently, from a conceptual standpoint, the just transition framework is grounded in three core dimensions of justice: distributive justice, procedural justice, and recognitional justice. Distributive justice addresses the fairness of outcomes—namely, how the benefits and burdens of the energy transition are allocated across social groups, regions, and generations. Procedural justice emphasizes the inclusivity and transparency of decision-making processes, ensuring that affected communities, workers, and other stakeholders have meaningful opportunities to participate in shaping transition pathways. Recognitional justice pertains to the acknowledgment of diverse identities, experiences, and structural disadvantages that may affect individuals' and groups' capacity to engage in or benefit from the transition.

As it was mentioned before, these dimensions are interdependent. For example, distributive outcomes often reflect the quality of procedural mechanisms, while recognition of social and cultural contexts is necessary to design inclusive participatory approaches. Consequently, the effectiveness of just transition policies depends not only on financial redistribution or employment programs but also on institutions' capacity to engage with affected populations in a meaningful, context-sensitive manner.

Institutionalization of the just transition framework has accelerated in recent years. Internationally, the Paris Agreement under the United Nations Framework Convention on Climate Change acknowledges the need for a just transition of the workforce. At the European level, the Green Deal integrates just transition principles through dedicated mechanisms such as the Just Transition Fund, aimed at supporting regions with high dependency on fossil fuels. These developments indicate a shift in climate governance paradigms toward more inclusive and equity-centered models of policy implementation.

Nevertheless, operationalizing the just transition framework presents several challenges. One of the most significant is the heterogeneity of transition impacts across sectors, regions, and population groups. For example, regions dependent on coal or heavy industry may face concentrated job losses, reduced fiscal revenues, and demographic decline, while urban centers may benefit disproportionately from green growth and innovation. Additionally, differences in national capacities, institutional quality, and labor-market structures shape the feasibility of just transition measures, especially in contexts with limited administrative resources or weak social protection systems.

Another challenge lies in managing temporal trade-offs. Many just transition measures, such as worker reskilling or regional economic diversification, yield benefits over the long term but require upfront investment and political commitment in the short term. Balancing immediate socioeconomic needs with long-term sustainability goals is therefore central to effective just transition strategies. In this context, transitional governance must be capable of coordinating multiple policy domains—energy, labor, education, social welfare—under a coherent framework that anticipates and mitigates unintended consequences.

Furthermore, the just transition must address not only employment and economic outcomes but also broader aspects of social cohesion, such as community identity, intergenerational equity, and democratic legitimacy. The decline of public trust in institutions, often exacerbated by perceived injustices or exclusion from decision-making, can undermine the social acceptability of low-carbon policies. Therefore, embedding justice principles into all phases of the energy transition—design, implementation, monitoring, and revision—is crucial to ensure both policy effectiveness and public support.

To summarize, the just transition framework serves as a strategic and normative guide for aligning decarbonization efforts with social equity imperatives. It recognizes that the benefits of climate action cannot be maximized, nor its costs minimized, without addressing the structural inequalities that shape who gains and who loses from the transition. By integrating distributive, procedural, and recognitional justice dimensions into climate and energy policy, the framework promotes a more resilient, inclusive, and socially sustainable transformation. As such, the just transition is not a parallel agenda to the energy transition—it is an essential condition for its success.

3.2 Distributional Effects of the Energy Transition

The global shift toward clean and low-carbon energy systems has accelerated significantly in recent years, driven by the urgent need to mitigate climate change and reduce environmental degradation. However, alongside these technological and environmental advances, questions of social equity have emerged as a critical area of concern. While the energy transition is often framed in terms of technological progress and emissions reductions, its socioeconomic implications—particularly the uneven distribution of benefits and burdens—deserve thorough consideration.

Rooted in the broader framework of social justice, the principle of energy equity affirms that all individuals should have access to affordable, reliable, and sustainable energy services. Such access is essential not only for fulfilling basic human needs but also for supporting livelihoods, reducing poverty, and enhancing quality of life. Yet, the distributional dynamics of the energy transition reveal that these benefits are not uniformly experienced. In many cases, vulnerable populations, marginalized communities, and less-developed regions face disproportionate challenges, including reduced affordability, limited infrastructural investment, and exclusion from emerging opportunities such as green employment or community-energy ownership.

Moreover, in the context of ongoing geopolitical instability, market fluctuations, and the aftermath of the COVID-19 pandemic, many governments have recalibrated their priorities toward short-term energy security. This shift often entails the continued reliance on fossil fuels and the deferral of structural reforms aimed at both decarbonization and social inclusion. As a consequence, the equity dimension of energy policy risks being pushed aside potentially exacerbating existing socioeconomic disparities and undermining the legitimacy and effectiveness of the transition itself.

Distributional effects of the energy transition encompass both positive and negative effects, reflecting how the costs, benefits, risks, and opportunities of systemic change are allocated across different social groups. These effects are dependent on the design and implementation of policies, as well as the governance structures that shape the energy sector, ultimately influencing the degree to which social justice is either advanced or undermined.

Positive distributional impacts emerge when energy transition benefits—including expanded access to affordable, sustainable energy services, new employment avenues, and improved social indicators such as health and education—are equitably distributed, particularly among marginalized and vulnerable populations. For instance, renewable energy interventions can play a pivotal role in mitigating energy poverty and reducing gendered labor burdens, thereby contributing to enhanced social inclusion.

Conversely, negative distributional consequences arise when existing structural inequalities are perpetuated or exacerbated by the uneven sharing of transition costs and benefits. Affordability barriers, inequitable access to technologies, displacement due to resource or land appropriation, and the reinforcement of gendered or racial disparities exemplify such outcomes.

3.2.1 Positive Distributional Effects

While challenges related to equity, access, and inclusivity remain salient, evidence increasingly suggests that—when carefully governed—the transition to clean energy systems can generate substantial socioeconomic gains. These gains are not limited to environmental improvements or climate resilience, but extend to tangible enhancements in living conditions, economic empowerment, and social cohesion. Recognizing the multi-level nature of these outcomes, the following sections explore the principal positive effects of the energy transition at the individual (micro), community (meso), and societal (macro) levels, highlighting the potential of well-designed energy policies to function as catalysts for inclusive development and long-term social sustainability.

It is worth noting that at both the micro and meso levels, the primary positive effects of the energy transition are closely associated with the tangible deployment of renewable energy infrastructure. For example, at the micro level, such interventions can significantly improve energy access, reduce household energy expenditures, and enhance overall quality of life, particularly in remote or underserved areas. They also enable greater energy self-sufficiency, which can contribute to long-term resilience and empowerment. At the meso level, the presence of renewable infrastructure often acts as a catalyst for local economic revitalization. It stimulates job creation, supports the development of local skills and entrepreneurship, and strengthens social cohesion through cooperative or community-based ownership models. These effects tend to be more immediate and visible, as they are embedded in localized contexts where renewable technologies intersect with everyday social and economic activity.

In contrast, macro-level impacts are less dependent on infrastructure deployment alone and are instead shaped by the broader policy and institutional environment that governs the energy transition (Lippert and Sareen, 2023). These effects—such as reductions in national poverty levels, structural shifts in employment across economic sectors, narrowing of regional disparities, and progress toward social equity—emerge more gradually and are subject to a complex interplay of regulatory frameworks, market dynamics, international cooperation, and fiscal instruments. For example, the extent to which the energy transition supports innovation, fosters technological competitiveness, or strengthens democratic governance at the national level depends on the alignment of climate, environmental, and energy policies with inclusive and long-term development strategies. As such, while the micro and meso levels reflect the immediate social dividends of renewable energy deployment, the realization of broader structural benefits at the macro level hinges on strategic policy coherence and robust governance mechanisms.

At the micro level, the energy transition brings a variety of positive socioeconomic outcomes that directly influence the everyday lives of individuals and households. These effects are particularly important, as they reflect how systemic shifts in the energy sector translate into tangible improvements at the level of personal experience and daily functioning.

- **Increased energy self-sufficiency.** The transition to decentralized renewable energy systems enables households to move from passive consumers to active energy producers, particularly through technologies such as rooftop solar photovoltaic (PV) panels, small-scale wind turbines, and home-based battery storage. This transformation enhances individual and household-level control over energy supply, increasing energy autonomy and reducing dependence on centralized grids that may be unreliable, especially in rural and remote areas. Energy self-sufficiency also empowers households to participate more actively in local energy markets, sometimes through mechanisms like net metering or energy cooperatives, which further contributes to energy democracy and resilience in the face of energy price volatility or geopolitical disruptions.
- **Reduction in energy expenditures.** Access to modern renewable energy technologies and improvements in energy efficiency can significantly reduce household energy expenditures over the medium and long term. While the initial investment cost of some technologies may be high, many are supported by public subsidies, concessional financing, or pay-as-you-go schemes that improve affordability. Once operational, renewable energy systems such as solar home systems or efficient cookstoves drastically reduce or eliminate recurring fuel costs (e.g., for kerosene, coal, or wood), allowing households to reallocate spending towards health, education, nutrition, or savings. The reduction in energy poverty through decreased household energy burdens is particularly impactful in low-income or rural contexts.
- **Positive employment effects.** The expansion of distributed renewable energy systems and energy efficiency interventions stimulates demand for labor-intensive services at the local level, including system installation, operation, maintenance, and technical training. These new employment opportunities are often more inclusive than those in centralized fossil-fuel industries, as they can be embedded within communities and tailored to local labor markets.
- **Improvements in daily life quality and overall well-being.** Reliable and clean energy access contributes to substantial improvements in the quality of daily life. It enables the use of electric lighting, refrigeration, communication devices, and various household appliances that increase comfort, and allow for greater convenience. In previously un-electrified or under-electrified settings, these improvements can transform daily routines, reduce stress associated with energy scarcity, and enable greater participation in economic and social life. The psychological benefits of energy security—such as increased feelings of safety, comfort, and dignity—are increasingly recognized as important contributors to subjective well-being.
- **Improvements in health conditions.** The replacement of polluting fuels such as firewood, charcoal, and coal with clean energy alternatives reduces household exposure to harmful air pollutants, including fine particulate

matter (PM2.5), carbon monoxide, and volatile organic compounds. This is particularly critical for women and children, who often spend more time indoors and are disproportionately affected by indoor air pollution. Health improvements include reduced incidence of respiratory and cardiovascular diseases, eye irritation, and burns, as well as long-term reductions in morbidity and mortality. Access to electricity also improves health outcomes by enabling cold chain technologies for vaccine storage, powering health clinics, and improving water purification and sanitation systems at the household level.

- **Reduction of women's domestic workload.** The energy transition has the potential to alleviate gendered burdens related to household labor. Access to clean energy technologies such as electric stoves, water pumps, washing machines, and lighting reduces the time and physical effort required for domestic tasks. In many low-income and rural contexts, women are primarily responsible for fuel collection, cooking, and water retrieval—activities that are not only time-consuming but also expose them to health and safety risks. By reducing the time spent on these tasks, energy access frees women's time for income-generating activities, education, or rest, thereby supporting gender equality and the empowerment of women.

  Sen et al. (2025) examined how women's empowerment influences the energy transition within the broader framework of sustainable development across developing countries. Drawing on panel data from 71 nations between 2001 and 2021, the study found that improvements in women's empowerment are positively associated with a range of energy transition indicators. Specifically, the analysis revealed a 9.03% increase in access to electricity, a 2.0% rise in the use of clean cooking fuels, a 20.8% growth in the share of renewable electricity, and a 2.9% improvement in energy efficiency. These findings underscore the multifaceted role of gender empowerment in shaping progress toward sustainable energy outcomes and offer valuable insights for policymakers committed to inclusive and equitable energy transitions.

- **Improved access to educational and learning conditions.** Electricity access enhances educational opportunities at multiple levels. At home, electric lighting enables children and adults to study during evening hours, while access to media and communication technologies broadens exposure to information, digital learning, and remote education. At schools, energy access supports the use of computers, internet connectivity, audiovisual learning tools, and climate control, all of which contribute to a more effective and inclusive learning environment. In rural and underserved areas, electrification has been positively associated with school attendance, academic performance, and the retention of qualified teachers—thus reinforcing human capital development in the long run.

At the meso level, the energy transition produces a set of transformative effects that manifest within communities, municipalities, and local institutions. These impacts go beyond individual outcomes, shaping the socioeconomic fabric of localities and influencing collective well-being.

- **New employment places and opportunities.** At the meso level, the deployment of renewable energy projects often serves as a direct source of employment creation, particularly in communities hosting new infrastructure. These employment effects span multiple phases—development, installation, operation, and decommissioning—each requiring varying levels of skills and qualifications. While high-skilled positions in design and engineering may be filled by external professionals, many medium- and low-skilled roles in construction, maintenance, and logistics are sourced locally. This creates a tangible economic benefit for host communities. Moreover, job creation linked to renewable energy often occurs in regions historically dependent on declining fossil-fuel industries, offering pathways for regional regeneration and labor-market revitalization. However, the net employment effect depends on the scale of investment, policy incentives, and the existence of complementary training programs to ensure local labor-market readiness.
- **Stimulation of local economic activity.** Beyond direct employment, the energy transition acts as a lever for broader local economic development. Clean energy projects tend to generate indirect and induced economic impacts through value chain linkages. Local businesses—such as equipment suppliers, transportation firms, service providers, and hospitality sectors—benefit from increased demand. Additionally, renewable energy production can lower energy costs and enhance supply reliability for small and medium-sized enterprises (SMEs), thus improving their competitiveness. Importantly, these dynamics can contribute to a more diversified and resilient economic structure at the local level, particularly in remote or structurally weak regions where economic activities are often limited. Economic stimulation also depends on procurement strategies: when local content requirements and inclusive procurement practices are adopted, the benefits are more likely to be retained within the community.
- **Local entrepreneurship and skills development.** The decentralized and modular nature of many renewable energy technologies—such as rooftop solar PV, small-scale wind, or bioenergy systems—creates significant opportunities for local entrepreneurship. These systems enable the establishment of community-owned energy cooperatives, municipal utilities, or small private enterprises, facilitating local ownership, revenue retention, and reinvestment into the community. Such models not only strengthen community agency in managing energy resources but also foster long-term, place-based economic resilience.

Simultaneously, the deployment and maintenance of these technologies necessitate the development of new skill sets. Vocational training programs, reskilling initiatives, and technical education—often implemented through public–private and civil-society partnerships—are critical to ensuring that local populations can effectively participate in and benefit from the energy transition.

Beyond the energy sector itself, the rise in household incomes—resulting from job creation, cost savings, or land lease revenues—can stimulate demand for a variety of local goods and services. This indirect economic multiplier effect leads to broader entrepreneurial activity and skill development across other sectors such as retail, transportation, construction, and various services. As communities experience rising economic activity, new markets emerge, necessitating additional competencies in areas such as management, finance, logistics, and customer service. Therefore, skills development linked to the energy transition is both sector-specific and cross-sectoral, reinforcing the importance of integrated local development strategies.

- **Strengthening of community cohesion.** Community-energy projects often require active participation from residents, municipalities, and civil-society actors. This participatory approach fosters dialogue, trust, and cooperation among stakeholders. Shared ownership models and inclusive decision-making can increase transparency and ensure that benefits—such as revenue, energy savings, or infrastructure improvements—are distributed fairly. These dynamics can lead to greater social cohesion and empowerment, especially when communities experience a sense of pride and collective identity in co-producing clean energy. Furthermore, the collaborative governance of energy projects may serve as a platform to build institutional capacity and foster civic engagement beyond the energy sector itself.
- **Enhanced community resilience.** Resilience at the community level is increasingly critical in light of climate-induced disruptions, energy price volatility, and global supply chain uncertainties. Decentralized renewable energy systems—such as microgrids or hybrid mini-grids—enhance energy autonomy and reliability, reducing dependence on centralized fossil-fuel-based systems. This is particularly relevant in remote or island communities where energy security is often fragile. Moreover, the integration of flexible technologies such as Virtual Power Plants (VPPs) further strengthens local resilience by enabling real-time load balancing, demand response, and full energy self-sufficiency, even during peak demand or external supply disruptions. These systems not only ensure uninterrupted access to electricity but also create adaptive energy ecosystems capable of responding dynamically to both environmental shocks and market fluctuations (Wang et al., 2023). In addition to technical resilience, clean energy infrastructure supports broader socioeconomic resilience by stabilizing energy access for schools, clinics, and essential services during crises. Moreover, the integration of community resilience strategies within

energy transition planning can enhance adaptive capacity and long-term sustainability.

- **Improved public services.** Clean energy investments frequently lead to tangible improvements in public service delivery. Electrification of educational institutions, health centers, and public administration facilities enables higher service quality and operational efficiency. Municipal services such as street lighting, water supply, and public transport also benefit from energy transition projects, especially when supported by smart technologies and digital innovation. Additionally, public revenues generated through taxes or leasing of municipal land for renewable projects can be reinvested into local infrastructure and services, reinforcing the developmental impact of the energy transition.

At the macro level, the energy transition drives also wide-ranging structural changes with significant implications for national economies, societal organization, and global development trajectories. The adoption of low-carbon energy systems contributes not only to climate mitigation but also to long-term socio-economic transformation.

- **Poverty alleviation.** At the macro level, the energy transition has the potential to substantially reduce poverty by enhancing access to affordable, clean, and reliable energy services, particularly in historically underserved or marginalized regions. The expansion of renewable energy infrastructure—such as mini-grids and off-grid solar systems—can provide essential energy access in remote and rural areas where centralized fossil-based systems have failed to reach. Increased access to modern energy sources facilitates income-generating activities, supports the development of small and medium-sized enterprises, and improves access to essential services like education, healthcare, and information technologies. By reducing the time and income spent on procuring traditional fuels and enabling more productive use of energy, the transition can contribute to the breaking of structural poverty traps. When aligned with redistributive energy policies, such as targeted subsidies or tariff reform, these benefits can be particularly significant for low-income populations.

  A growing body of literature highlights the beneficial role of low-carbon energy transitions in addressing energy poverty. For instance, Bonatz et al. (2019) conducted a comparative analysis examining the relationship between energy poverty and low-carbon development in Germany and China. Their findings suggest that numerous policy instruments aimed at low-carbon development—particularly those focused on improving energy efficiency and expanding renewable energy—are closely tied to alleviating energy poverty. By lowering overall energy consumption and enhancing access to modern, reliable energy sources, such measures contribute to improved energy affordability and quality, especially for vulnerable populations.

- **Employment places and labor-market transformation.** One of the most widely acknowledged macroeconomic benefits of the energy transition is its potential to generate new employment opportunities across multiple sectors. The expansion of renewable energy industries—ranging from solar PV and wind energy to energy efficiency services and grid modernization—has already proven to be a significant source of job creation. An illustrative example is China, where the deployment of renewable energy infrastructure has progressed at an exceptionally rapid pace. Li et al. (2021) investigated the impact of energy conservation and emissions reduction policies on employment dynamics in China over the period of 2010 to 2013. The findings demonstrated that such policies have a significant positive effect on net employment growth, primarily through enhancing job creation and mitigating job destruction within firms. Specifically, a 1% increase in the intensity of these policies was associated with an 8.04% rise in net employment. Further heterogeneity analysis revealed that the positive employment effects were less pronounced in firms with high export intensity, elevated market concentration, and high pollution levels. In contrast, firms operating in regions with a higher degree of market orientation experienced stronger gains in net job creation, suggesting that institutional and market conditions play a critical role in shaping the employment outcomes of low-carbon policy interventions. Other research indicates that in 2010, a 1% increase in the share of solar PV generation in China was associated with a 0.68% rise in total employment—representing the highest employment elasticity among all power generation technologies (Cai et al., 2011).

  The deployment of renewable energy technologies contributes not only to the creation of new employment opportunities but also induces structural changes in the labor market. As the sector expands, there is a growing demand for a skilled workforce capable of supporting the installation, operation, and maintenance of increasingly sophisticated technologies (Li et al., 2023; Liu et al., 2023). This shift promotes the development of new professional profiles and encourages the upskilling and retraining of workers, particularly those transitioning from traditional fossil-fuel-based sectors. In this way, the energy transition serves as a catalyst for long-term human capital development, fostering a more resilient and adaptive labor force. Nevertheless, recent research highlights that the employment gains associated with economic decarbonization are distributed unevenly across regions, sectors, and skill levels, reflecting the differentiated absorptive capacities of local labor markets (Muttitt and Gass, 2023).
- **Reduction of regional disparities.** The spatial dynamics of renewable energy deployment can contribute significantly to reducing regional inequalities. Unlike fossil-fuel infrastructure, which tends to be centralized and capital-intensive, renewable energy projects can be implemented in decentralized forms and are often located in rural or economically

disadvantaged areas rich in natural resources (Tachega et al., 2025). These projects can stimulate regional economies through direct employment, local procurement, and infrastructure development. Additionally, they improve access to public goods. By attracting investment and enabling economic diversification in lagging regions, the energy transition can help rebalance national development patterns. Governments can amplify these effects by adopting regionally differentiated energy policies and directing clean energy subsidies or incentives toward structurally weaker areas.

- **Advancing gender equality.** At the macro level, the energy transition offers a structural pathway to mitigate long-standing gender inequalities embedded within traditional energy systems. In many regions, women are disproportionately affected by energy poverty, spending significant time on unpaid domestic labor such as fuel collection and food preparation using inefficient and polluting energy sources. These burdens not only limit their participation in education and formal employment but also expose them to severe health risks, particularly from indoor air pollution. The introduction of electrification and access to clean cooking technologies can significantly alleviate these constraints—reducing time poverty, improving health outcomes, and expanding opportunities for women to engage in economic and social life. Moreover, macro-level policy instruments, including national energy strategies, regulatory frameworks, and institutional reforms, play a critical role in embedding gender considerations into energy governance. By mainstreaming gender equity and actively including women's perspectives in decision-making processes, the energy transition can contribute to more inclusive, equitable, and socially just development outcomes. The study by Ying et al. (2025) provides empirical evidence that greater representation of women in political institutions, alongside higher levels of patent-driven innovation, are positively associated with the uptake of renewable energy technologies. This relationship remains consistent across multiple quantiles and temporal scales, suggesting that both gender-inclusive governance and technological advancement serve as enabling factors in the energy transition.
- **Promotion of social justice and distributional equity.** The energy transition provides an opportunity to embed principles of social justice into the architecture of national energy systems. At the macro scale, this requires addressing the historical and structural inequalities embedded in energy access, affordability, and decision-making. Policy instruments such as progressive pricing schemes, public ownership models, participatory planning processes, and redistributive taxation can help ensure that the benefits of the transition are equitably shared while mitigating its potential harms. For example, funds raised through carbon pricing can be redistributed through social transfers or used to support communities undergoing industrial restructuring. Institutionalizing the concept of a "just transition" in national policy frameworks not only fosters social

cohesion but also enhances political legitimacy and public support for ambitious climate and energy goals.

- **Support for climate and sustainability goals.** From a macroeconomic perspective, the energy transition is not only a cornerstone for achieving long-term climate stability and environmental sustainability but also a critical mechanism for advancing social equity and justice. The decarbonization of the energy sector remains essential to fulfilling global climate commitments—particularly those enshrined in the Paris Agreement—which necessitate substantial reductions in greenhouse-gas (GHG) emissions across all major economies. The accelerated deployment of renewable energy technologies and energy efficiency improvements contributes not only to reduced GHG emissions, but also to the mitigation of local air and water pollution and the deceleration of ecosystem degradation, thereby lowering the incidence of climate-related health risks and economic disruptions.
- Crucially, when energy transition strategies are aligned with broader national development frameworks, they can promote policy coherence and enhance the potential to achieve multiple, interlinked goals. This includes environmental targets such as biodiversity protection and sustainable land-use management, as well as social objectives aimed at addressing systemic inequalities. Integrated transition policies can facilitate more equitable access to clean and affordable energy services, reduce regional and socioeconomic disparities, and support inclusive governance structures that ensure meaningful participation of marginalized and underrepresented groups. As such, the energy transition—if designed with intentionality—serves not only as an environmental imperative but as a transformative opportunity to embed principles of fairness, inclusion, and justice into the foundation of a sustainable future.
- **Stimulation of innovation and green industrial development.** The shift toward a low-carbon economy is inherently linked to processes of industrial and technological innovation. Public and private investments in renewable energy Research and Development and Demonstration (RD&D) are fostering rapid technological advancements in solar PV, battery storage, smart grids, and hydrogen, among others. These innovations reduce the cost of clean technologies and create new markets, enabling countries to position themselves as leaders in the green economy. Strategic industrial policies that prioritize domestic value chains—through localization requirements, green procurement, and public investment—can further enhance national competitiveness and technological sovereignty. Over time, this fosters the emergence of green clusters, supports knowledge-intensive growth, and reduces dependence on imported technologies.

However, existing empirical evidence suggests that public investments in research, development, and demonstration (RD&D) for renewable energy technologies may exert upward pressure on household electricity prices, at least in the short term. For instance, Macedo and Marques

(2023) find that while RD&D efforts targeting fossil-fuel technologies slightly reduce electricity prices—attributable to the technological maturity and cost-efficiency of these systems—investments in RD&D for renewable energy tend to increase household electricity costs. This effect is primarily linked to the substantial upfront expenditures, innovation risks, and infrastructural integration challenges associated with emerging renewable technologies. These findings underscore the need for carefully balanced RD&D strategies that support innovation while also considering the affordability dimension, particularly for vulnerable energy consumers.

- **Energy security and strategic autonomy.** A just energy transition, driven by the expansion of renewable energy systems, enhances national energy security by reducing reliance on fossil-fuel imports, which are often subject to geopolitical tensions and price volatility (Vasic et al., 2023; Zhang et al., 2023). The deployment of clean technologies increases the overall resilience of energy systems to external disruptions, natural disasters, and global market shocks. This is especially critical in the context of recent global crises—such as the COVID-19 pandemic and the war in Ukraine—which have exposed the vulnerabilities of international energy supply chains (Siksnelyte-Butkiene, 2023). By diversifying the energy portfolio and promoting domestic, low-carbon energy production, the transition not only strengthens national strategic autonomy but also contributes to long-term economic stability and sustainability.
- **Positive fiscal and trade balance effects.** Reduced import of fossil fuels and the growth of domestic clean energy industries contribute to improved fiscal and trade balances (Idris et al., 2023). Lower energy import bills free up public and private resources for reinvestment in e.g., social infrastructure, education, or industrial development. Simultaneously, countries that develop competitive renewable energy and energy efficiency technologies can benefit from green exports, strengthening their trade position in the global economy. When integrated with broader industrial and export strategies, the energy transition can become a key pillar of national economic resilience.
- **Strengthening of democratic governance and institutional transparency.** The decentralized structure of renewable energy systems creates space for more participatory and transparent governance models. Community-owned energy projects, prosumer arrangements, and local decision-making structures allow for increased citizen engagement in energy planning. This shift can help to rebalance power dynamics in the energy sector, moving away from monopolistic governance structures. At the macro level, enhanced transparency and democratic accountability in energy decision-making increase institutional legitimacy and policy stability, both of which are critical for long-term energy transition success (Healy and Barry, 2017).

To provide a comprehensive overview of the socioeconomic outcomes associated with the shift to low-carbon energy systems, Table 3.1 outlines the main positive distributional effects of the energy transition across three levels of analysis: micro, meso, and macro. This categorization allows for a clearer assessment of the distributional dimensions of the energy transition and helps identify the scale at which different policy interventions may be most effective.

Overall, it can be stated that he energy transition generates a wide range of positive impacts at the micro (individual and household) level, contributing directly to improvements in social welfare and quality of life. Among the most notable effects are enhanced health conditions due to reduced exposure to air pollutants, increased energy self-sufficiency, and lower energy-related expenditures. In particular, the integration of decentralized renewable energy systems enables households to gain greater control over their energy use and reduce vulnerability to price volatility. The transition also contributes to a reduction in the domestic burden traditionally borne by women, particularly in regions reliant on biomass or manual energy sources, thereby supporting gender equality. Improvements in educational outcomes are frequently observed as well, especially when enhanced electricity access supports digital learning and lighting availability. In parallel, the expansion of clean energy technologies fosters

Table 3.1 The main positive distributional effects of the energy transition at multiple levels. Source: developed by the author.

Micro level (individual and household)
• Increased energy self-sufficiency
• Reduction in energy expenditures
• Creation of new employment opportunities
• Improvements of daily life quality and overall well-being
• Improvements in health conditions
• Reduction of women's domestic workload
• Improved access to educational and learning conditions
Meso level (community)
• New employment places and opportunities
• Stimulation of local economic activity
• Local entrepreneurship and skills development
• Strengthening of community cohesion
• Enhanced community resilience
• Improved public services
Macro level (societal)
• Poverty alleviation
• Positive employment effects
• Reduction of regional disparities
• Reduction of gap among genders
• Promotion of social justice and equity
• Support for climate and sustainability goals
• Acceleration of clean technology innovation and green industrial development
• Strengthening democratic governance
• Improvement of energy security
• Positive fiscal and trade balance effects

new employment opportunities, generating both direct and indirect socio-economic benefits.

In addition to these primary effects, a number of complementary benefits have been identified in the literature, particularly in specific regional and socio-economic contexts. For instance, the deployment of renewable energy systems in off-grid or underserved areas enhances access to basic services and fosters digital inclusion, particularly through improved connectivity and information access. Transition-related developments also contribute to greater individual empowerment, particularly where participatory models allow households to act as prosumers or community-energy stakeholders (Siksnelyte-Butkiene et al., 2023a). Moreover, the replacement of hazardous or polluting traditional fuels with modern energy alternatives improves household safety and reduces the risk of fire and indoor air pollution-related illnesses. Finally, by supporting more stable and diversified household livelihoods, the energy transition contributes to long-term intergenerational benefits in health, education, and economic security.

At the meso level, the energy transition fosters significant socioeconomic benefits within communities and local economies. The most prominent impacts include the creation of new employment opportunities and the stimulation of local economic activity, which together contribute to regional development. Support for entrepreneurship and the development of relevant skills further enhance the capacity of communities to adapt and benefit from emerging clean energy sectors. Additionally, the transition strengthens social cohesion by bringing stakeholders together around shared goals, enhancing community resilience in the face of environmental and economic changes. Improvements in public services, often linked to new energy infrastructure and increased local revenues, further contribute to the quality of life and well-being of community members. Collectively, these effects support a more inclusive and sustainable development trajectory at the community level.

Beyond these primary outcomes, the energy transition generates a range of complementary benefits that reinforce its socioeconomic impacts at the meso scale. Increased local participation in energy planning and decision-making processes promotes community empowerment and enhances the legitimacy and social acceptance of renewable energy projects. This participatory dynamic can also foster stronger networks and social capital, enabling communities to better manage change and leverage new opportunities. Infrastructure improvements, including better roads, communication, and energy distribution systems, often accompany renewable energy projects and provide broader developmental spillovers. Moreover, the transition encourages technological innovation and knowledge diffusion within local economies, facilitating diversification and enhancing regional competitiveness. By reducing environmental burdens such as air and water pollution, the energy transition improves local ecosystem services, which are critical for agriculture, recreation, and cultural identity. Taken together, these additional impacts contribute to building the adaptive capacity and long-term resilience of communities navigating the energy transition.

At the macro level, the energy transition offers a range of structurally significant and interrelated benefits that contribute to broader socioeconomic and environmental objectives. Among the most notable are the alleviation of poverty through improved access to modern, affordable, and reliable energy services and the net creation of employment opportunities across clean energy value chains. The transition also plays a critical role in reducing regional disparities by stimulating investment and economic activity in less-developed or structurally disadvantaged areas. Moreover, when supported by inclusive policies, it contributes to narrowing gender gaps and enhancing social inclusion, thereby reinforcing principles of equity and social justice. From an environmental governance perspective, the shift toward low-carbon energy systems supports the fulfillment of national and international climate commitments and broader sustainability goals. In parallel, the transition fosters innovation and accelerates technological advancement in renewable energy, digital infrastructure, and green industries, strengthening long-term competitiveness. Enhancing energy security through reduced dependence on imported fossil fuels and promoting more resilient, transparent, and participatory governance structures further underscore the strategic relevance of the energy transition at the national and global scale.

3.2.2 Negative Distributional Effects

While the energy transition is often associated with opportunities for environmental and socioeconomic improvement, growing evidence indicates that its distributional outcomes are not universally beneficial. An increasing number of scientific publications analyze the social implications of the energy transition, focusing on the effects of low-carbon energy technologies, climate change mitigation policies, and specific energy policy measures on society and social equality. These studies reveal that, while the shift towards sustainable energy systems creates opportunities to reduce social disparities, it may also exacerbate existing inequalities if social aspects are not sufficiently integrated into policy planning and implementation. This subsection examines the negative distributional effects of the energy transition, discussing how certain groups bear disproportionate costs or are systematically excluded from the benefits of clean energy developments.

In recent years, the deployment of renewable energy technologies has expanded significantly, reflecting their increasing role in meeting global energy demands and contributing to the achievement of international climate targets. As renewable energy initiatives are central element to shift towards low-carbon energy transition, their impact on social equity can be discussed first. There is a growing body of scientific research addressing this topic. One of the most comprehensive attempts to synthesize the findings of various case studies conducted globally is the study by Johnson et al. (2020). Johnson et al. (2020) conducted a systematic review of the scientific literature examining how low-carbon energy technologies influences gender and social equity. Their analysis identifies several interrelated consequences for society. First, the dual impact on women's labor conditions, where both empowerment and increased burdens can occur was identified. Second, the persistence of poverty, employment insecurity, and precarious livelihoods issues were singled out. Third, the loss of land and the continuation of colonial patterns driven by market-oriented interests were pointed out. Finally, the persistent inequities in access to the benefits of energy transitions, which reflects deeper systemic exclusions was underscored.

- **Dual impact on women's labor conditions.** Modern forms of household energy—particularly electricity for lighting and clean technologies for cooking and heating—are often identified as instrumental in promoting gender equality in rural areas. Decentralized solar solutions, such as small PV systems, are especially valuable for low-income households or living in remote areas. By extending available hours for educational and productive activities, these technologies have been associated with improved learning outcomes for children, enhanced health, and increased livelihood opportunities.

  Nevertheless, the analysis of Johnson et al. (2020) showed that a growing body of empirical research suggests that such interventions do not uniform deal with existing social hierarchies. Rather than eradicating structural inequalities, energy transitions can, in some cases, merely reconfigure or displace them. This phenomenon is particularly evident in the reorganization of gendered labor patterns. While renewable energy technologies often reduce the time women spend on tasks such as firewood collection and traditional cooking, this time savings does not necessarily translate into personal empowerment. In many contexts, women's responsibilities are merely redirected toward other unpaid or underrecognized forms of domestic or agricultural labor.

  Moreover, the authors found that the distribution of benefits from energy access frequently reflects entrenched gender norms and socioeconomic stratification. For example, men are often better positioned to leverage newly available energy for entrepreneurial or recreational purposes, while women tend to allocate additional time to household duties or income-generating activities driven by necessity rather than choice. The scope for women to capitalize on new energy opportunities is further constrained by limited access to markets, credit, and ownership of commercial infrastructure—barriers that are reinforced by prevailing gender ideologies.

  Additionally, the intersection of gender and class amplifies disparities in who can effectively benefit from clean energy investments. In communities where economic power and social mobility are unevenly distributed, it is typically the better-off and male-headed households that are able to convert energy access into tangible improvements in well-being or economic productivity. As such, the transformative potential of renewable energy is contingent not only on technological deployment but also on the broader social and institutional environment (Johnson et al., 2020).

- **The persistence of poverty, employment insecurity, and precarious livelihoods.** The integration of renewable energy systems into low-income and rural communities is recognized as a pathway to poverty alleviation and improved livelihoods. However, according to the analysis performed by Johnson et al. (2020) some empirical evidence revealed that the outcomes of such interventions are far from uniform. While renewable energy deployment can generate new employment opportunities, in some cases these may be characterized by informal, low-wage, and precarious conditions, especially in large-scale bioenergy or plantation-based systems. In such contexts, land-use changes and contractual arrangements often disadvantage small producers and destroy traditional livelihoods, resulting in limited or even negative socioeconomic impacts for vulnerable populations.

  Moreover, the assumption that renewable energy inherently fosters social equity overlooks the complex interplay between gender, class, and local socioeconomic dynamics. Gendered divisions of labor persist even after energy access improves, with benefits unevenly distributed and often reinforcing existing inequalities. Women, in particular, may face increased labor burdens or reduced access to livelihood resources when energy transitions are introduced without addressing entrenched social norms.

  It shows that renewable energy, if implemented without structural and institutional safeguards, may not solve the challenges of poverty and precarity in the energy transition.
- **The loss of land and the continuation of colonial patterns driven by market-oriented interests.** In some cases renewable energy development can be framed in ways that marginalize indigenous, rural, and vulnerable communities by prioritizing broader environmental objectives over social interests of local people. A significant concern in the scientific literature are expressed towards land loss, especially linked to large-scale renewable energy developments such as biofuel plantations and hydropower projects. These developments frequently reveal pronounced conflicts between interests of private sector and local people, with the loss of land often resulting in diminished livelihoods and food insecurity (Nsafon et al., 2023).

  The displacement of community rights is evident across various renewable energy initiatives globally, including large-scale solar, wind, biofuel, and hydropower projects. Issues of land tenure and ownership are central to the challenges faced by affected communities, where benefits from renewable energy accrue disproportionately to external investors. However, it necessary to emphasize that most of such examples are evident in developing or least developed countries, suffering from lack of democracy, corruption, lack of social policies, and so on.

  The repercussions of land loss and forced resettlement extend beyond economic deprivation, deeply altering the social structures within affected communities. This includes shifts in power dynamics and entrenched gender roles. Research highlights differentiated impacts on men and women, particularly where loss of access to land and traditional livelihood activities leads to increased vulnerability. Displaced populations are frequently relocated to areas unsuitable for agriculture, pushing many into informal labor markets or precarious economic activities. Research showed that men generally face fewer barriers in securing such employment, whereas women often experience heightened economic dependency, reinforcing existing gender inequalities in the post-transition context.

The research indicates that changes in traditional livelihoods caused by land loss can lead to increased outmigration. This phenomenon is also linked to gender issues, as women often face challenges securing employment in new locations and may experience a greater physical labor burden when their husbands migrate to urban areas in search of work (Ding et al., 2019).

- **Inequities in access to the benefits of energy transitions.** According the research performed by Johnson et al. (2020), despite that numerous studies highlight that renewable energy projects, while often heralded for their potential to foster rural development and reduce gendered labor burdens, can simultaneously reinforce existing social and economic inequalities through uneven access. Solar energy, for example, has improved energy self-sufficiency and educational outcomes in various rural contexts. However, financial barriers—such as high upfront costs for household systems—often exclude the poorest households, perpetuating cycles of debt and deepening disparities (Ahlborg, 2017). The gendered dimension of these barriers is particularly notable; women, who frequently have less control over household income and property, face greater challenges in accessing solar technologies. Community-driven, women-led initiatives in some villages have demonstrated how participatory approaches can facilitate more inclusive energy access, leading to shifts not only in energy-use patterns but also in local gender norms and women's social standing.

  Also, the scientific studies showed that in some cases subsidy schemes do not fully alleviate the financial burden, leaving poor households unable to afford high down payments and ongoing fees. This has resulted in subsidies disproportionately benefiting wealthier households, while poorer families remain marginalized. Furthermore, in some cases subsidies have widened socioeconomic divides, reinforcing unequal access to the benefits of solar electrification, including enhanced education, health, and communication. Such challenges advocate for more socially differentiated financial mechanisms that incorporate the needs and insights of local communities.

  In higher-income settings, studies reveal that the expansion of residential solar PV has produced uneven distributional effects. For instance, in Germany, rapid growth in solar capacity has been accompanied by rising electricity prices, disproportionately impacting lower-income households through regressive cost burdens. Similarly, in the United States, solar PV adoption varies significantly along racial lines, with Black and Hispanic communities having markedly lower rooftop solar installation rates (Johnson et al., 2020). This disparity correlates with differences in income and property ownership, underscoring how delayed participation in renewable energy deployment risks exacerbating preexisting socioeconomic inequalities.

A portion of the literature provides evidences on assessing the social impacts associated with one or several specific energy sources. While renewables offer considerable environmental and energy security benefits, their integration into existing landscapes introduces visible and often irreversible changes to natural environment. As renewable energy installations—such as wind turbines, solar farms, hydro energy, and bioenergy facilities—replace or modify landscape features, public perceptions of these transformations become a critical factor influencing the overall social acceptability of such projects (Cohen et al., 2014; Spiess et al., 2015; Manyoky et al., 2016; Lennon and Scott, 2017; Vuichard et al., 2021). The transition toward a low-carbon energy system is not only altering physical landscapes but is also reshaping the sociocultural meaning attached to those spaces (Siksnelyte-Butkiene et al., 2023b). This transformation will persist as decarbonization efforts intensify. In this context, as emphasized by Drozdz et al. (2021), actively involving local communities in the planning and decision-making processes is essential to enhance procedural fairness and foster greater societal acceptance of emerging energy technologies. Building trust, ensuring transparency, and respecting local values are therefore indispensable components of a just and inclusive energy transition. However, many studies have demonstrated that, although the public generally holds favorable views toward renewable energy and acknowledges its role in mitigating climate change, the potential for experiencing negative externalities—particularly changes in the landscape—often leads to opposition or resistance to specific projects (Dugstad et al., 2020; Oehlmann et al., 2021). Within these investigations, additional negative consequences such as landscape changes, diminished quality of life attributable to noise, visual disturbances, or impact to the environment have been consistently observed. Such findings highlight the multifaceted nature of social effects rising from energy infrastructure development, underscoring the need for comprehensive impact assessments.

- **Degradation of the environment/loss of ecosystem services.** A substantial body of research has highlighted various environmental impacts resulting from the implementation of new energy infrastructure. Low-carbon energy technologies typically require extensive land areas for the deployment of generation facilities and supporting infrastructure, such as transmission networks. This spatial demand results in significant modifications to landscape elements (Pang et al., 2014; Aziz, 2023), which in turn can lead to disruptions in ecosystem services. For instance, wind farms change the visual character of the landscape and generate noise, which may affect both human well-being and local wildlife. Hydropower installations can induce even more profound physical transformations by modifying river courses, fragmenting habitats, and impacting ecological integrity, thereby compromising a range of ecosystem services. Similarly, utility-scale solar PV systems necessitate large areas of land, often in areas that host diverse flora and fauna, potentially threatening habitat availability and biodiversity. For example, Fraser and Chapman (2018) conducted an in-depth assessment of the social consequences associated with the development of large-scale solar power installations in Japan. While their analysis highlighted a range of beneficial outcomes, the study also identified significant adverse social effects, notably the risk of environmental degradation linked to these projects.

Therefore, as a result, local communities frequently express opposition not only due to anticipated landscape transformation but also in response to perceived or actual degradation of ecosystem services. Although fossil-fuel-based technologies also exert spatial and environmental pressures, renewable energy systems are often characterized by a larger land footprint and lower energy density, necessitating greater land use per unit of energy produced (Pasqualetti and Stremke, 2018). The transition to renewable energy also requires the extensive use of specific scarce metals such as copper, lithium, cobalt, and nickel (Henckens, 2022). These spatial and ecological considerations highlight the importance of integrated spatial planning that balances the technical requirements of low-carbon energy systems with the preservation of environmental and social values.

- **Changes in the landscape.** The visual dimension of renewable energy technologies and their associated infrastructure plays a central role in shaping how energy-scape elements are integrated into existing landscapes, particularly within urbanized environments. A substantial body of scientific work has examined public responses to various renewable energy technologies, with a predominant focus on aesthetic perceptions (e.g., Devine-Wright and Batel, 2013; Valtchanov and Ellard, 2015; Salak et al., 2021). Among the identified factors, aesthetic considerations often emerge as the most immediate and pronounced. For instance, Betakova et al. (2015) reported that landscapes characterized by a high density of technical devices are generally perceived as being of lower visual quality compared to those with minimal infrastructure presence. Similarly, findings by Spielhofer et al. (2021a) demonstrated that people exhibit higher levels of physiological arousal when observing landscapes dominated by energy infrastructure, and consistently express a preference for more natural, less visually occupied environment. Further research by Spielhofer et al. (2021b) underscored the importance of aesthetic coherence and the perceived visual harmony between energy devices and the surrounding landscape, identifying these elements as key determinants of public acceptance. While aesthetic values are clearly influential, the planning and deployment of renewable energy infrastructure should also account for a broader spectrum of ecosystem services, including cultural, provisioning, and regulating functions, to ensure a more holistic and socially sustainable integration.

The incorporation of energy-related infrastructure into the landscape—often referred to as energy-scape—can significantly transform ecosystem services that underpin essential human needs, including health, food provision, and opportunities for recreation (Siksnelyte-Butkiene et al., 2023b). These transformations may affect how local communities perceive and interact with their environment. Interestingly, empirical evidence suggests an inverse relationship between proximity to planned energy infrastructure and public acceptance; individuals residing at greater distances from proposed renewable energy developments tend to exhibit higher levels of acceptance (Meyerhoff et al., 2010).

The duality between positive and negative outcomes underscores the complexity of evaluating the societal implications of renewable energy infrastructure, where positive advancements may coexist with unintended negative externalities. Therefore, it is important to pay attention that despite the fact that renewable energy initiatives are central to the low-carbon transition, they are not inherently equipped to resolve persistent social inequities. These interventions often unfold within complex sociocultural and economic landscapes where structural disparities—such as unequal access to resources, services, and decision-making power—remain deeply embedded. Without explicitly addressing these underlying inequalities, renewable energy development risks entrenching the very patterns of marginalization it seeks to overcome. When energy systems are restructured without confronting power imbalances, particularly those linked to class, gender, ethnicity, and geography, new regimes may reproduce the same old structure (Johnson et al., 2020). Therefore, for the energy transition to be really inclusive and transformative, it must go beyond technical solutions and integrate a critical analysis of how energy is governed, who benefits, and whose voices are heard in shaping the future energy landscape.

Also, the scientific literature provides research-based evidence on the social justice implications regarding income inequality, employment rate, and structural unemployment of climate policy.

- **Exacerbation of preexisting socioeconomic disparities.** Scientific research highlights that when climate change mitigation and just transition policies do not sufficiently incorporate targeted measures addressing the needs of marginalized population groups and do not prioritize the principles of social justice, they risk reinforcing existing socioeconomic disparities rather than reducing them. For example, Fragkos et al. (2021) conducted a comprehensive assessment of the socioeconomic effects arising from the implementation of emission reduction targets in the EU. Their findings indicate that achieving climate neutrality may lead to a modest increase in income inequality, particularly affecting lower-income households. Nevertheless, the adverse impacts can be mitigated through the strategic redistribution of carbon tax revenues. When these revenues are allocated as lump-sum transfers to households or used to lower social security contributions, such measures can enhance employment levels and contribute to reducing income disparities among EU citizens. Semet (2024) explored the socioeconomic implications associated with the implementation of carbon taxation in France. The findings underscore a paradox: although the carbon tax proves to be an effective instrument in reduce household-level emissions, its distributional consequences raise significant equity concerns. The policy disproportionately affects lower-income households, who experience a tax burden nearly three times higher than that of wealthier households. Moreover, the study reveals that a considerable share of the emissions reduction achieved to the tax is offset by increased emissions from overcompensated households. These compensatory measures, rather than mitigating inequality, contribute to further disparities in the tax burden both within and across income groups.

- **Job losses in traditional fossil-fuels based industry**. In countries where fossil-fuels constitute the predominant share of the energy mix, the transition toward cleaner energy systems is frequently accompanied by significant employment disruptions, particularly within conventional energy industries. Although the expansion of renewable energy and other low-carbon technologies generates new employment opportunities, a direct occupational transfer for workers previously engaged in fossil-fuel-based sectors is seldom feasible (Li et al., 2023; Liu et al., 2023; Greenspon and Raimi, 2024). This is largely due to differences in required skill sets, geographic location of new jobs, and the structural mismatch between declining and emerging industries.

  China offers a salient example of this dynamic. As the world's largest producer and consumer of coal, the country faces considerable challenges in advancing its low-carbon transition. The coal sector remains not only a critical component of the national energy portfolio but also a major source of employment (Huang et al., 2019), particularly in certain provinces where economic development is heavily dependent on mining activities. This dual reliance—both economic and demographic—complicates efforts to shift away from coal without triggering substantial social and regional dislocations. Zhang et al. (2024) explored the relationship between the supply-side energy transition and employment by employing panel data from China spanning 2008 to 2022. Their analysis revealed that a 1% increase in clean energy generation was associated with an approximate 0.013% rise in overall employment. However, the impact varied across sectors. Employment in the primary sector declined, whereas the tertiary sector experienced growth. From an economic perspective, less-developed regions saw greater benefits, suggesting a potential to reduce talent disparities. In terms of resource endowments, coal-rich regions faced notable job losses, underscoring the importance of a just transition. In contrast, areas with abundant wind and solar resources recorded stronger employment growth, highlighting the scale effects of clean energy expansion. However, the effects are strongly influenced by the type of renewable energy technology, the structure of financing mechanisms for renewable energy subsidies, and the extent and definition of employment impact considered (Mu et al., 2018).

In addition to the structural barriers highlighted above, numerous case studies underscore the negative socioeconomic impacts of the energy transition on marginalized groups and displaced workers. For instance, Luke (2023) investigated the labor market consequences of the clean energy shift in the United States and found persistent policy gaps in addressing job losses in carbon-intensive industries. The study also draws attention to the historical exclusion of women and people of color from energy sector employment, and emphasizes the increasing economic precarity of dislocated workers in the absence of targeted social protections and inclusive policy frameworks. These findings further reinforce the necessity of embedding equity considerations within just transition strategies.

Thus, China's energy transition underscores the importance of integrating just transition frameworks that address employment retraining, social protection, and regional economic diversification in parallel with climate mitigation efforts.

The transition towards sustainable energy systems, while essential for mitigating climate change, generates a range of negative distributional effects that manifest differently across social scales. By analyzing these impacts through a multi-level framework—comprising the micro (individual and household), meso (community), and macro (societal) levels—Table 3.2 identifies the principal adverse consequences associated with the energy transition. Understanding these effects is critical to informing equitable policy design and ensuring that the benefits of the energy transition do not exacerbate existing social and economic disparities.

At the micro level, the energy transition exerts a dual impact on women's labor conditions. Although improvements in household energy services, such as cleaner cooking and more efficient heating, reduce time devoted to these specific tasks, women frequently reallocate this time towards increased agricultural work, livestock management, and other subsistence activities, effectively intensifying their overall workload. Concurrently, households often confront heightened poverty and employment insecurity, as job displacement in traditional fossil-fuel sectors is insufficiently compensated by new opportunities in low-carbon energy projects, which may be unstable. Livelihood loss at this level also encompasses loss of land, which undermines the economic base of many households and contributes to increased vulnerability. This combination of factors drives patterns of outmigration for work, frequently towards urban centers or other cities, thereby disrupting established family and social

Table 3.2 The main negative distributional effects of the energy transition at multiple levels. Source: developed by the author

Micro level (individual and household)
• Dual impact on women's labor conditions
• Poverty and employment insecurity
• Job losses in traditional fossil fuel industries
• Livelihood loss
• Outmigration
• Inequitable access to benefits
Meso level (community)
• Loss of land/displacement
• Food insecurity
• Disrupted livelihoods
Macro level (societal)
• Reinforcement of existing social and economic inequalities
• Increase in disparities among genders
• Enlarged economic disparities
• Structural unemployment
• Exclusion of marginalized groups
• Loss of cultural identity and heritage
• Changes in landscape and environmental degradation

networks. Furthermore, access to the benefits of the energy transition remains inequitable: many families are unable to afford the necessary technologies or cover initial installation costs, lack access to loans, or are unable to use subsidy programs, reinforcing existing socioeconomic disparities.

At the meso level, the energy transition manifests through spatial and socioeconomic transformations within communities. Large-scale renewable energy developments often lead to land loss and displacement, disproportionately affecting populations dependent on traditional land tenure systems. Such changes can compromise local food security by reducing the availability of arable land and disrupting established agricultural practices. As a result, community livelihoods face significant disruption, particularly in contexts where institutional mechanisms for compensation, adaptation, or participatory planning remain weak or inaccessible. Moreover, newly introduced energy projects may lack long-term stability or integration into broader regional development strategies, generating economic uncertainty and reducing the resilience of affected communities.

At the macro level, the energy transition may reinforce and widen structural social and economic inequalities. The distribution of costs and benefits across regions and population groups may remain uneven, with wealthier actors and urban populations more likely to capture the gains associated with renewable energy development, while disadvantaged groups may bear a disproportionate share of the burdens. Gender disparities may be amplified, as women often face limited access to financial instruments—such as loans or subsidies—necessary for adopting clean technologies. Moreover, women remain underrepresented in energy-related industries and decision-making processes, which restricts their influence over transition pathways and limits their access to the socioeconomic benefits generated by the transition. Furthermore, prevailing gender norms and stereotypes continue to assign domestic responsibilities to women. As a result, the time and labor savings from cleaner household energy services are frequently redirected toward other physically demanding tasks, such as agricultural or informal work, rather than contributing to rest, education, or professional advancement. Additionally, the transformation of energy systems and the reallocation of land for renewable energy infrastructure may undermine the cultural identity and heritage of indigenous populations, particularly where these shifts disrupt long-standing economic structures and traditional land use. These dynamics risk deepening social fragmentation and generating resistance to energy reforms, particularly in regions already marked by socioeconomic vulnerability.

It is also important to acknowledge that social impacts associated with energy transitions are highly context-dependent and may vary significantly based on a wide array of influencing factors. These include, but are not limited to, the type of energy technology deployed, the design and implementation of policy instruments, the scale and pace of project development, and the geographical, socioeconomic, and cultural characteristics of the affected area. Moreover, the level of public awareness, community literacy, and access to information play a critical role in shaping perceptions and responses to energy projects. Additional considerations such as historical land-use patterns, preexisting socioeconomic disparities, and local governance structures further complicate the social dimension of energy transitions. Consequently, making generalized or definitive claims regarding the social effects of low-carbon energy policies at large risks oversimplifying a complex and multidimensional process. A nuanced, place-based approach is therefore essential for accurately assessing the diverse and evolving social outcomes of the energy transition.

3.3 Intergenerational Equity in Energy Transitions

The discourse on social justice within the energy sector increasingly emphasizes the necessity of incorporating intergenerational considerations into policy frameworks. While much of the current literature and policy discourse focuses on inequalities among present populations, the concept of intergenerational equity addresses the ethical and policy-related obligation to preserve the capacity of future generations to meet their own energy and environmental needs. This principle is particularly pertinent given the long-term externalities of energy production and use, including GHG emissions, resource depletion, and environmental degradation.

Energy systems and infrastructure are inherently long-lived and often carry environmental and economic outcomes for decades. Policy decisions made today regarding energy production, investment, and consumption have the potential to significantly affect future generations, especially in terms of climate stability, ecosystem integrity, and access to energy services. The reliance on carbon-intensive energy sources, while contributing to short-term economic and social objectives, imposes substantial long-term risks related to environmental degradation and climate change. These impacts disproportionately affect populations not yet born, raising critical concerns regarding temporal justice and ethical responsibility.

Furthermore, the exclusion of future generations from current decision-making processes introduces an asymmetry in rights and representation. As such, present-day societies bear a normative responsibility to act as stewards of finite environmental and energy resources. Integrating intergenerational equity into energy policy thus requires the adoption of planning horizons that extend beyond investment cycles.

Therefore, ensuring intergenerational equity in energy policy is essential for achieving sustainable development goals and maintaining the integrity of environmental and energy systems over the long term. Addressing this issue of equity calls for both ethical reflection and the implementation of forward-looking, evidence-based policy instruments that account for the welfare of both current and future generations.

3.3.1 *The Social Discount Rate and Long-term Policy Assessment*

The large-scale consumption and production of energy remain among the principal contributors to environmental degradation and climate change, generating significant implications not only for present populations but also for future generations. Addressing contemporary energy and environmental challenges thus necessitates a forward-looking approach that integrates the ethical dimensions of intergenerational responsibility. In this regard, climate change stands as a salient example of a policy domain where intergenerational equity is paramount. A fundamental question emerges: to what extent do current generations bear moral and practical obligations to avoid actions that may result in long-term environmental harm, thereby compromising the welfare of those not yet born? The assessment of these obligations plays a central role in shaping climate mitigation strategies and the scale at which emissions reductions should be pursued over time. In economic literature, this is frequently conceptualized through the lens of the social discount rate, which determines the weight assigned to future benefits and costs in present decision-making.

The social discount rate plays an important role in the assessment of long-term public policy decisions, especially in areas such as climate change mitigation, energy policy, infrastructure development, or intergenerational justice. It functions as an essential instrument for converting future economic costs and benefits into present values, thereby enabling a consistent comparison of short-term investments with long-term societal outcomes. As emphasized in scientific discourse, the choice of an appropriate social discount rate has profound implications for sustainability and equity across generations. Mathematically, the social discount rate is often represented by the Ramsey formula (Ramsey, 1928). In the Ramsey framework, it is assumed that future utility should be discounted at a fixed rate (δ), reflecting a preference for present consumption. The discount rate formula is expressed as:

$$r = \delta + \eta * g$$

In this formula, r denotes the social discount rate, δ reflects the rate of pure time preference, η captures the elasticity of marginal utility with respect to consumption, and g indicates the anticipated long-term rate of economic growth.

The pure rate of time preference (δ) reflects the notion that individuals may prefer immediate benefits over future ones, even if the outcomes are identical. This tendency may arise from uncertainty about the future, perceived risks, or intrinsic preferences for immediacy. In parallel, the component η * g incorporates anticipated economic growth, acknowledging that future generations are likely to enjoy higher income levels; thus, each additional unit of consumption would bring them comparatively lower marginal utility. This approach ensures that the social discount rate reflects diminishing marginal utility, aligning with the principle that the value of consumption decreases as wealth increases.

Pure time discounting, in particular, entails reducing the value of future outcomes solely based on their temporal distance from the present, independently of changes in wealth or uncertainty. The pragmatic component of the social discount rate is often reflected through the inclusion of a pure time preference parameter. This element reflects the normative assumption that benefits or well-being occurring in the present are inherently more valuable than identical outcomes realized in the future. As a result, it implies that the life and experiences of current generations are assigned a higher intrinsic value than those of individuals born in the future. Embedded in the Ramsey framework, this practice assumes that future utility should be discounted at a fixed rate (δ), reflecting a preference for present consumption. In the context of climate policy, the choice of δ significantly influences the evaluation of long-term environmental impacts. A higher pure rate of time preference reduces the present value of future climate damages, potentially leading to less aggressive mitigation efforts. Conversely, a lower or zero δ places greater importance on future outcomes, supporting more substantial immediate actions to address climate change.

The application of pure time discounting in economic evaluations of long-term public policies—particularly in climate change mitigation and sustainability assessments—has attracted significant criticism. The core concerns can be summarized as follows:

- Ethical considerations. One of the most prominent criticisms is ethical in nature. As first articulated by Ramsey (1928), the use of a positive pure rate of time preference (δ) suggests that future utility is inherently less valuable than present utility, which he deemed "ethically indefensible." This principle implies that the lives and well-being of future generations hold less intrinsic value simply due to their temporal position, thereby contradicting the fundamental notions of intergenerational justice.

- Intergenerational equity. Applying a pure time discount rate in climate policy assessments leads to systemic undervaluation of long-term damages, thereby transferring disproportionate environmental burdens to future generations. This approach is inconsistent with the principles of intergenerational equity and the just transition framework, which demand equitable treatment of all generations regardless of when they live.
- Contradiction with sustainability objectives. A high pure rate of time preference results in present-biased policies, undermining commitments to long-term environmental sustainability. This contradicts international sustainability frameworks, including the UN SDGs, which promote the safeguarding of planetary and human systems for future generations.
- Arbitrariness of parameter selection. There is significant inconsistency in the empirical or normative justification for the choice of δ. The value of the pure time preference rate is often set without transparent criteria, leading to divergent policy recommendations.
- Philosophical and theoretical inconsistencies. From a philosophical standpoint, many scholars argue that discounting future utility solely due to its temporal location represents a form of unjustifiable temporal discrimination. Therefore, applying δ undermines the coherence of normative ethical frameworks that assign equal moral worth across generations.

Growth-based discounting is an economic principle, positing that future generations are expected to be wealthier than the current one. This approach, grounded in the utilitarian principle of diminishing marginal utility of consumption, justifies placing a lower weight on future damages since wealthier future societies are presumed to be better equipped to absorb the associated costs. As it was mentioned before, this assumption embedded in the Ramsey formula through the $\eta * g$ term. As some economists explain, sustained economic growth in industrialized countries over the past century underpins this rationale, whereby climate mitigation costs should be distributed progressively across generations according to their income levels (Tarsney, 2017). Consequently, the welfare of future generations is discounted on the premise that they will be wealthier and thus better positioned to bear the costs of mitigation.

However, such an approach raises critical concerns, particularly when assessed from the perspective of sustainability and intergenerational equity. First, projections of sustained economic growth are inherently uncertain in the context of climate change, environmental degradation, and geopolitical instability (Stern, 2007). Overreliance on optimistic growth scenarios may underestimate the risks facing future populations, including potential declines in well-being and ecosystem services. Second, growth discounting implies that the welfare of future individuals is valued less solely due to their position in time, which may contradict ethical principles of distributive justice. This creates an imbalance where the present generation prioritizes its short-term economic interests while transferring disproportionate environmental burdens to future generations, thereby undermining the principles of a just transition and sustainable development (Caney, 2014).

Moreover, economic metrics such as GDP fail to capture the full scope of human well-being and environmental quality. Even if future generations experience economic growth, they may suffer from reduced natural capital, social instability, or compromised health—factors not adequately reflected in consumption-based discounting frameworks. In light of these concerns, some scientists have doubts and criticism about social discounting approach, indicating that this approach is useless and misleading from both: the ethical and economic perspectives. Therefore alternative evaluative frameworks are needed that give proper weight to the rights and interests of future populations.

3.3.2 Application of Social Discount Rates in Practice

Social discount rates differ considerably across countries, shaped by economic structure, developmental imperatives, and normative perspectives on intergenerational justice. Broadly, high-income economies tend to adopt lower discount rates—typically within the range of 1% to 6%—in recognition of ethical arguments regarding future welfare and the declining marginal utility of consumption over time. In contrast, many developing economies employ higher rates, often between 7% and 15%, to reflect elevated capital costs, higher market interest rates, and urgent developmental needs.

The selection of an appropriate SDR is not uniform even within countries; it often varies depending on the sectoral context, temporal scope of the project, and specific policy priorities. For instance, in the United States, the Office of Management and Budget revised its long-standing guidance in November 2023. It replaced the dual-rate approach—where a 3% rate reflected the social rate of time preference and a 7% rate represented the opportunity cost of capital—with a unified real discount rate of 2% (The Office of Management and Budget. Circular, 2023). This adjustment, embedded in updates to Circular A-4, marks a substantial methodological shift. It underscores a growing recognition of the need to incorporate long-term benefits and intergenerational equity into regulatory impact assessments, particularly in domains such as climate policy and public health.

Another illustrative case from the global economy is the EU. According to the European Commission's framework for cost-benefit analysis of investment projects co-financed by EU structural and investment funds, a real discount rate of 5% is advised for Cohesion Member States—typically those with gross national incomes below the EU average—while a lower rate of 3% is suggested for more developed Member States (European Commission, 2015). Though formulated for EU-funded projects, these benchmarks reflect the broader range of social discount rates applied across the Union, generally oscillating between 3% and 5%.

In the case of China, there is also no standardized social discount rates applied across all projects. Nevertheless, empirical studies suggest that discount rates generally range between 5% and 8%, depending on the project's characteristics. Lower rates are typically applied to projects with long-term environmental or social returns, while higher rates are employed for infrastructure or energy-related investments to reflect capital opportunity costs and associated investment risks.

However, ethical considerations play a pivotal role, when considering intergenerational equity, particularly in the context of evaluating long-term impacts such as those associated with climate change, environmental degradation, and just transition. While individual economic agents often exhibit time preferences that favor present consumption over future gains, societal decision-making—especially in the public sector—must not uncritically mirror this bias. The application of temporal discounting, if not properly justified, risks undervaluing the welfare and rights of future generations.

Future individuals, unlike present stakeholders, are inherently excluded from current democratic and institutional processes, rendering them voiceless in decisions that will profoundly affect their quality of life. This asymmetry raises fundamental ethical concerns regarding justice, fairness, and responsibility. The principle of intergenerational equity requires that current generations recognize their role as stewards of natural, economic, and social capital, ensuring that future populations inherit a world that allows for equal or improved opportunities for well-being.

Accordingly, the selection of a social discount rate must reflect not only economic parameters but also normative considerations regarding the weight assigned to future benefits and harms. As such, purely utilitarian frameworks may be insufficient in capturing the full moral dimensions of long-term public investment decisions, particularly those with irreversible consequences for human and ecological systems.

In summary, the application of traditional discounting methods, particularly those based on Ramsey-style discounting, has notable limitations in addressing climate change impacts. This approach fails to adequately capture the wide range of climate-related risks that go beyond economic consumption. The most significant impacts of climate change—such as the degradation of irreplaceable environmental goods, public health crises, displacement, and the potential for existential threats—are non-marginal, non-substitutable, and do not diminish in significance over time as economic consumption increases. While future technological advancements, such as geoengineering or other mitigation strategies, may offset some of these effects, the discount rate does not appropriately account for the pace, scale, or uncertainty surrounding such innovations. Thus, evaluating climate change costs and benefits purely through the lens of discounted future consumption can lead to underestimating the true extent of climate risks.

References

Ahlborg, H. Towards a conceptualization of power in energy transitions. Environmental Innovation and Societal Transitions, 2017, 25, pp. 122–141.

Aziz, T. Accounting impacts of renewable energy expansions on ecosystem services to balance the trade-offs. Science of the Total Environment, 2023, 879, 162990.

Betakova, V.; Vojar, J.; Sklenicka, P. Wind turbines location: how many and how far? Applied Energy, 2015, 151, pp. 23–31.

Bonatz, N.; Guo, R.; Wu, W.H.; Liu, L.J. A comparative study of the interlinkages between energy poverty and low carbon development in China and Germany by developing an energy poverty index. Energy & Buildings, 2019, 183, 817–831.

Cai, W.J.; Wang, C.; Chen, J.N.; Wang, S.Q. Green economy and green jobs: Myth or reality? The case of China's power generation sector. Energy, 2011, 36(10), 5994–6003.

Caney, S. Climate change, intergenerational equity and the social discount rate. Politics Philosophy & Economics, 2014, 13(4), pp. 320–342.

Carley, S.; Konisky, D.M. The justice and equity implications of the clean energy transition. Nat. Energy, 2020, 5(8), pp. 569–577.

Cohen, J.J.; Reichl, J.; Schmidthaler, M. Re-focussing research efforts on the public acceptance of energy infrastructure: A critical review. Energy, 2014, 76, pp. 4–9.

Devine-Wright, P.; Batel, S. Explaining public preferences for high voltage pylon designs: an empirical study of perceived fit in a rural landscape. Land Use Policy, 2013, 31, pp. 640–649.

Ding, W.; He, L.; Zewudie, D.; Zhang, H.; Zafar, T.B.; Liu, X. Gender and renewable energy study in Tibetan pastoral areas of China, Renewable Energy, 2019, 133, pp. 901–913.

Drozdz, W.; Kinelski, G.; Czarnecka, M.; Wójcik-Jurkiewicz, M.; Maroušková, A.; Zych, G. Determinants of Decarbonization—How to Realize Sustainable and Low Carbon Cities? Energies 2021, 14, 2640.

Dugstad, A.; Grimsrud, K.; Kipperberg, G.; Lindhjem, H.; Navrud, S. Acceptance of wind power development and exposure – not-in-anybody's-backyard. Energy Policy, 2020, 147, pp. 111780.

European Commission. European Commission. Guide to Cost-Benefit Analysis of Investment Projects Economic appraisal tool for Cohesion Policy 2014–2020. 2015, 364 p. Brussels, Belgium. Available online at: https://ec.europa.eu/regional_policy/sources/studies/cba_guide.pdf [Assessed on 09-05-2025].

Fragkos, P.; Fragkiadakis, K.; Sovacool, B.; Paroussos, L.; Vrontisi, Z.; Charalampidis, I. Equity implications of climate policy: Assessing the social and distributional impacts of emission reduction targets in the European Union. Energy, 2021, 237, 121591.

Fraser , T .; Chapman , A . J . Social equity impacts in Japan's mega-solar siting process. Energy for Sustainable Development, 2018, 42, 136–151.

Greenspon, J.; Raimi, D. Matching geographies and job skills in the energy transition. Extractive Industries and Society, 2024, 17, 101397.

Healy, N.; Barry, J. Politicizing energy justice and energy system transitions: Fossil fuel divestment and a "just transition". Energy Policy, 2017, 108, pp. 451–459.

Henckens, M.L.C.M. The Energy Transition and Energy Equity: A Compatible Combination? Sustainability, 2022, 4781.

Henry, M.S., Bazilian, M.D., Markuson, C. Just transitions: Histories and futures in a post-COVID world. Energy Research & Social Science, 2020, 68, 101668.

Huang, H.; Roland-Holst, D.; Springer, C.; Lin, J.; Cai, W.J.; Wang, C. Emissions trading systems and social equity: A CGE assessment for China. Applied Energy, 2019, 235, 1254–1265.
Idris, F.M.; Seraj, M.; Özdeser, H. Renewable energy consumption, CO2 emissions and trade balance nexus in OECD countries: evidence from ARDL bounds approach. International Journal of Energy Sector Management, 2023, 17(4), pp. 645–660.
Johnson, O.W.; Han, J.Y.C.; Knight, A.L.; Mortensen, S.; Aung, M.T.; Boyland, M.; Resurrección, B.P. Intersectionality and energy transitions: A review of gender, social equity and low-carbon energy. Energy Research & Social Science, 2020, 70, 101774.
Lennon, M.; Scott, M. Opportunity or Threat: Dissecting Tensions in a Post-Carbon Rural Transition. Sociologia Ruralis, 2017, 57(1), pp. 87–109.
Li, H.B.; Zhang, B.B.; Wen, L.; Li, Z. On the relationship between the energy conservation and emissions reduction policy and employment adjustment by manufacturing firms: Microdata from China. Journal of Cleaner Production, 2021, 297, 126652.
Li, Z.R.; Wang, M.; Wang, Q.W. Job destruction and creation: Labor reallocation entailed by the clean air action in China. China Economic Review, 2023, 79, 101945.
Lippert, I.; Sareen, S. Alleviation of energy poverty through transitions to low-carbon energy infrastructure. Energy Research & Social Science, 2023, 100, 103087.
Liu, G.; Chen, T.C.; Sui, X.; Solangi, Y.A. Examining and prioritizing the effect of sustainable energy on the job market to advance China's green workforce. Heliyon, 2023, 9(12), e22710.
Luke , N . Just Transition for All? Labor organizing in the energy sector beyond the loss of "jobs property." Annals of the American Association of Geographers, 2023, 113(1), 94–109.
Macedo, D.P.; Marques, A.C. Is the energy transition ready for declining budgets in RD&D for fossil fuels? Evidence from a panel of European countries. Journal of Cleaner Production, 2023, 417, 138102.
Manyoky, M.; Wissen Hayek, U.; Heutschi, K.; Pieren, R.; Gret-Regamey, A. Evaluating a visual-acoustic simulation for wind park assessment. Landscape and Urban Planning, 2016, 153, pp. 180–197.
McCauley, D.; Heffron, R. Just transition: Integrating climate, energy and environmental justice. Energy Policy, 2018, 119, pp. 1–7.
Meyerhoff, J.; Ohl, C.; Hartje, V. Landscape externalities from onshore wind power. Energy Policy, 2010, 38(1), pp. 82–92.
Mu, Y.Q.; Cai, W.J.; Evans, S.; Wang, C.; Roland-Holst, D. Employment impacts of renewable energy policies in China: A decomposition analysis based on a CGE modeling framework. Applied Energy, 2018, 210, 256–267.
Muttitt , G .; Gass , P . How energy transition affects jobs. Nature Climate Change, 2023, 13, 1162–1163.
Nsafon, B.E.K.; Same, N.N.; Yakub, A.O.; Chaulagain, D.; Kumar, N.M.; Huh, J.S. The justice and policy implications of clean energy transition in Africa. Frontiers in Environmental Science, 2023, 11, 1089391.
Oehlmann, M.; Glenk, K.; Lloyd-Smith, P.; Meyerhoff, J. Quantifying landscape externalities of renewable energy development: Implications of attribute cut-offs in choice experiments. Resource and Energy Economics, 2021, 65, 101240.
Okereke, C. Climate justice and the international regime. Wiley Interdisciplinary Reviews: Climate Change, 2010, 1(3), pp. 462–474.
Pang, X.; Mortberg, U.; Brown, N. Energy models from a strategic environmental assessment perspective in an EU context—what is missing concerning renewables? Renewable Sustainable Energy Reviews, 2014, 33, pp. 353–362.

Pasqualetti, M.; Stremke, S. Energy landscapes in a crowded world: A first typology of origins and expressions. Research in Social Sciences, 2018, 36, pp. 94–105.

Ramsey, F.P. A Mathematical Theory of Saving. The Economic Journal, 1928, 38(152), pp. 543–559. Available online: https://www.jstor.org/stable/pdf/2224098.pdf

Salak, B.; Lindberg, K.; Kienast, F.; Hunziker, M. How landscape-technology fit affects public evaluations of renewable energy infrastructure scenarios. A hybrid choice model. Renewable and Sustainable Energy Reviews, 2021, 143, 110896.

Schlosberg, D. Reconceiving environmental justice: global movements and political theories. Environmental Politics, 2004, 13 (3), pp. 517–540.

Semet , R . Coordinating social equity and emissions: Challenges in carbon tax policy. Energy Policy, 2024, 185, 113954.

Sen, K.K.; Karmaker, S.C.; Chapman, A.J.; Saha, B.B. Women's empowerment in driving the energy transition for sustainable development in developing nations. Renewable & Sustainable Energy Reviews, 2025, 216, 115647.

Siksnelyte-Butkiene I. Defining the Perception of Energy Security: An Overview. Economies. 2023, 11(7), pp. 174.

Siksnelyte-Butkiene, I.; Streimikiene, D.; Balezentis, T.; Volkov, A. Enablers and barriers for energy prosumption: Conceptual review and an integrated analysis of business models. Sustainable Energy Technologies and Assessments, 2023a, 57, 103163.

Siksnelyte-Butkiene, I.; Streimikiene, D.; Agnusdei, G., Paolo.; Balezentis, T. Energyscape concept for the transition to a low-carbon energy society. Environment Development and Sustainability, 2023b, 25, pp. 14953–14973.

Spielhofer, R.; Thrash, T.; Hayek, U.W.; Gret-Regamey, A.; Salak, B.; Grubel, J.; Schinazi, V.R. Physiological and behavioral reactions to renewable energy systems in various landscape types. Renewable & Sustainable Energy Reviews, 2021a, 135, 110410.

Spielhofer, R.; Hunziker, M.; Kienast, F.; Hayek, U.W. Does rated visual landscape quality match visual features? An analysis for renewable energy landscapes. Landscape and Urban Planning, 2021b, 209, 104000.

Spiess, H.; Lobsiger-Kagi, E.; Carabias-Hutter, V.; Marcolla, A. Future acceptance of wind energy production: Exploring future local acceptance of wind energy production in a Swiss alpine region. Technological Forecasting and Social Change, 2015, 101, pp. 263–274.

Stern N. The Economics of Climate Change: The Stern Review. Cambridge University Press; 2007.

Tachega, M.A.; Chen, Y.J.; Wang, J.J.; Yang, P. Regional disparities in Africa's renewable energy transition: Patterns and drivers. Journal of Cleaner Production, 2025, 503, 145396.

Tarsney, C. Does a Discount Rate Measure the Costs of Climate Change? Economics and Philosophy, 33, 2017, pp. 337–365.

The Office of Management and Budget. Circular No. A-4. 2023, Available online at: https://bidenwhitehouse.archives.gov/wp-content/uploads/2023/11/CircularA-4.pdf (Assessed on 09-05-2025).

Valtchanov, D.; Ellard, C.G. Cognitive and affective responses to natural scenes: effects of low level visual properties on preference, cognitive load and eye-movements. Journal of Environmental Psychology, 2015, 43, pp. 184–195.

Vasic, B; Pekic, I; Simic, G. Energy security of the European Union and corruption in Central Asia as the main challenges for the European sustainable energy future. Energy Sustainability and Society, 2023, 13, 1, p. 31.

Vuichard, P.; Stauch, A.; Wustenhagen, R. Keep it local and low-key: Social acceptance of alpine solar power projects. Renewable and Sustainable Energy Reviews, 2021, 138, 110516.

Wang, X.X.; Lo, K. Just transition: A conceptual review. Energy Research & Social Science, 2021, 82, 102291.

Wang, Y.F.; Gao, W.J.; Li, Y.; Qian, F.Y.; Yao, W.X. Techno-economic analysis of the transition toward the energy self-sufficiency community based on virtual power plant. Frontiers in Energy Research, 2023, 11, https://doi.org/10.3389/fenrg.2023.1010846

Williams, S.; Doyon, A. Justice in energy transitions. Environmental Innovation and Societal Transitions, 31 (2019), pp. 144–153.

Ying, W.; Ramzan, M.; Olanrewaju, V.O.; Brika, S.K.; Abozrib, A. The role of democratic governance and patent innovations in advancing sustainable energy transition. Energy & Environment, 2025, https://doi.org/10.1177/0958305X251343060

Zhang, B.; Liu, Z.Y.; Wang, Z.H. The impact of geopolitical risk on energy security: Evidence from a GMM panel VAR approach. Resources Policy, 2023a, 86, Part B, 104222.

Zhang, R.N.; Li, W.L.; Li, Y.H.; Li, H. Job losses or gains? The impact of supply-side energy transition on employment in China. Energy, 2024, 308, 132804.

CHAPTER 4

Energy Poverty and Social Equity

4.1 The Concept of Energy Poverty and Its Research Directions

Energy poverty remains one of the most pressing challenges in contemporary energy systems, directly affecting the well-being of households and the equitable distribution of energy services. Beyond the immediate economic constraints on access to energy, energy poverty reflects broader societal inequities, encompassing dimensions such as gender, age, ethnicity, disability, and geographic location. Understanding energy poverty is also increasingly relevant in the context of the global low-carbon transition. Policies promoting renewable energy deployment, energy efficiency, and decarbonization have the potential to either alleviate or exacerbate energy inequalities. Without careful design, technological advancements and market-based energy solutions may inadvertently deepen disparities in access, affordability, and participation. Therefore, assessing energy poverty through the lens of social equity is essential not only to protect vulnerable groups but also to ensure the sustainability, inclusiveness, and legitimacy of energy transitions.

The concept of energy poverty encompasses a range of interrelated socioeconomic and technical issues. It is generally understood as a condition in which households are unable to meet their basic energy needs. This deprivation often results from a combination of factors, including insufficient income, high energy prices, poor energy efficiency of buildings, reliance on outdated or inefficient technologies, and limited access to reliable or modern energy sources (Pye et al., 2015; Charlier and Kahouli, 2019; Streimikiene et al., 2020). Numerous studies have demonstrated that energy poverty is closely linked to negative health outcomes, including elevated mortality rates during colder seasons and declines in mental well-being (Liddell et al., 2015; Kose, 2019; Kahouli, 2020; Recalde et al., 2019).

I. Siksnelyte-Butkiene, D. Streimikiene, *Social Equity in the Energy Sector*,
https://doi.org/10.1007/978-3-032-22224-4_4

A broader interpretation of energy poverty has been proposed in global frameworks. For example, the United Nations Development Programme (2000) describes energy poverty as the inability to secure adequate, safe, affordable, high-quality, and environmentally sustainable energy. In recent years, scholarly attention has increasingly focused on how energy-poor households face systemic disadvantages—such as exclusion from competitive energy markets or limited access to energy-related services—that lead to disproportionately high energy expenditures and compound social inequalities (Dobbins et al., 2019; Ambrosio-Albala et al., 2020; Groves et al., 2020; Karakara et al., 2021).

Despite growing consensus on the significance of energy poverty, inconsistencies in terminology continue to pose challenges. Two closely related but distinct concepts are commonly used to describe limited access to adequate energy services. The term fuel poverty is primarily applied in contexts such as the United Kingdom and Ireland, where it refers to households that cannot afford sufficient energy for heating and other essential needs. This concept is typically assessed using indicators that capture the interaction between household income, energy prices, and the energy efficiency of dwellings (Bouzarovski and Petrova, 2015). Conversely, the term energy poverty is more widely employed across international contexts and refers to a broader set of issues, including overall inadequacy of energy services and systemic exclusion from energy infrastructure and governance (Halkos and Gkampoura, 2021).

Nonetheless, across both high-income and low-income countries, these concepts converge around a common core: the inability of individuals or households to access and afford a socially and materially acceptable level of energy services. Recognizing this condition as a fundamental aspect of inequality is essential for informing just and inclusive energy policy.

The issue of energy poverty is of relevance to both developed and developing economies; however, its manifestation and underlying aspects differ substantially depending on the context. In higher-income countries, discussions tend to focus on the proportion of household income devoted to meeting energy needs, as well as questions of energy affordability, energy vulnerability, and social justice. In contrast, in lower-income countries, energy poverty is primarily associated with limited or absent access to modern energy sources and infrastructure. Accordingly, the concept is approached from two dominant perspectives:

a. situations where a substantial share of income must be allocated to energy consumption and
b. an inability to obtain modern, reliable, and clean energy services.

In academic literature, energy poverty is explored through a variety of analytical lenses. One strand of research examines its implications for health and social aspects (e.g., Kahouli, 2020; Abbas et al., 2021a, 2021b; Oliveras et al., 2021; Mohan, 2021), whereas another focuses on economic determinants and outcomes (e.g., Kyprianou et al., 2019; Lin and Wang, 2020; Garba and

Bellingham, 2021). A further body of work considers the influence of policy and regulatory frameworks (Primc and Slabe-Erker, 2020; Uche-Soria and Rodríguez-Monroy, 2020). Moreover, the level of analysis varies widely across studies, ranging from household and firm-level assessments to national and regional examinations. Overall, the research directions in the field of energy poverty may be classified into several broad thematic groups (Fig. 4.1).

A considerable body of research has been dedicated to measuring energy poverty at the household level, reflecting widespread recognition of its importance for living standards and human well-being. Household-based assessments remain the most prevalent type of energy poverty measurement in both developed and developing countries. In recent years, studies have also begun to explore energy poverty among enterprises, acknowledging that insufficient access to energy services or excessively high energy costs can constrain productivity and development. Despite similarities in overarching objectives, the orientation of these assessments differs markedly depending on the level of socioeconomic development of a country. In developed countries, energy poverty research typically emphasizes social and distributive aspects, relying mainly on single indicators such as expenditure thresholds or housing characteristics. These analyses often draw on conceptual frameworks such as fuel poverty, energy vulnerability, and energy justice. In contrast, studies conducted in developing country contexts focus more closely on energy access, particularly

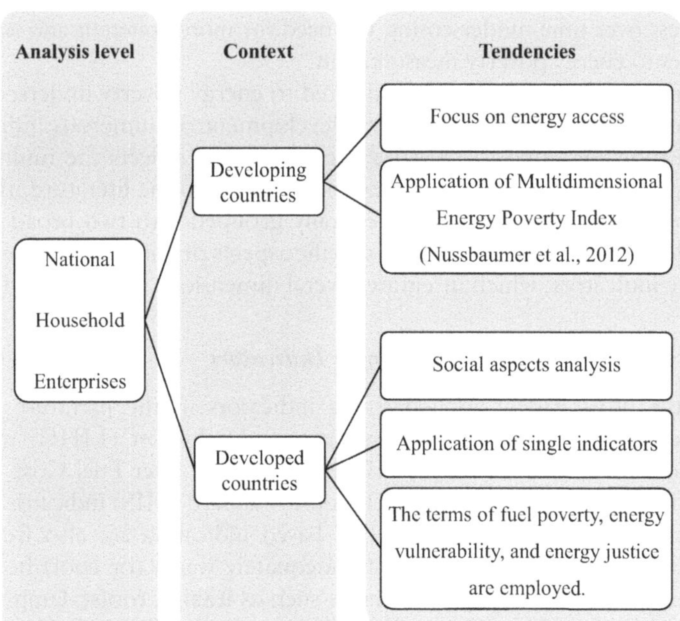

Fig. 4.1 The key thematic groups of energy poverty research. Source: developed by the author based on Siksnelyte-Butkiene et al. (2022).

the availability and reliability of modern energy services such as electricity and clean cooking fuels. To capture this broader development context, researchers frequently apply multidimensional indices, most notably the Multidimensional Energy Poverty Index proposed by Nussbaumer et al. (2012), which evaluates deficits across several energy service domains.

4.2 Energy Poverty Measurement

Although the measurement of energy poverty represents an essential starting point for designing effective policy interventions, there is currently no unified and widely accepted index. Instead, measurement approaches are highly fragmented and vary according to the level of evaluation, the developmental stage of the country, and the specific dimensions intended to be captured. As a result, indicator sets are assembled in a context-sensitive manner. In developed countries, measurement tools tend to emphasize affordability, household income–expenditure relationships, and housing energy efficiency, often relying on single or expenditure-based indicators. By contrast, in developing countries, greater attention is paid to access to modern energy services, availability of basic infrastructure, and reliability of supply, which favors the use of multidimensional indices. Additional variation stems from differences in policy priorities, climatic conditions, urban–rural composition, and the availability of statistical data, all of which shape the selection of indicators. The absence of a harmonized framework not only reduces comparability of results across countries and regions but also hampers the ability to monitor progress over time, underscoring the need for more coherent and adaptable approaches to energy poverty measurement.

The growing body of research dedicated to energy poverty underscores the continued relevance of this issue. The development of numerous indices and analytical tools designed to measure energy poverty reflects the multifaceted and complex nature of the phenomenon. In the scientific literature, the indicators used for assessment can be generally grouped into two broad categories: single indicators, which capture specific aspects of energy deprivation, and composite indicators, which integrate several dimensions into a unified metric.

4.2.1 Single Indicators

Among the most widely applied single indicators in the literature are the 10% expenditure threshold, the Low-Income High-Cost (LIHC) measure, the Twice the National Median (2M) indicator, the After Fuel Cost Poverty (AFCP) measure, and the Minimum Income Standard (MIS) indicator. Within the European context, three EU-SILC-based indicators are also frequently employed: (i) inability to maintain an adequately warm (or cool) home, (ii) presence of physical housing deficiencies such as leaking roofs, damp walls or floors, and window frame decay, and (iii) arrears on utility bills (Heindl and Schuessler, 2015; Siksnelyte-Butkiene et al., 2021). In practice, empirical studies often rely on a combination of several single indicators to provide a more robust assessment of energy poverty.

The 10 percent indicator, originally proposed by Boardman (1991), categorizes a household as energy poor when at least one-tenth of its income is devoted to energy expenditures. It is widely used due to its simplicity and ease of application. The LIHC measure, developed by Hills (2012), identifies energy poverty when two conditions are simultaneously met: a household spends more than the population median on energy, and its remaining income after energy costs falls below the official poverty line. Under the Minimum Income Standard (MIS) approach, a household is considered energy poor if, after paying for energy services, its income lies below a socially defined minimum acceptable standard of living (Moore, 2012). The 2M indicator comprises four related metrics—double the median, double the mean, double the median share, and double the mean share of household energy spending—and classifies households as energy poor if their outlays exceed these thresholds (Castaño-Rosa et al., 2019). The AFCP indicator identifies households that, after meeting housing and essential living expenses, lack sufficient income to afford necessary energy services (Hills, 2011). By contrast, the EU-SILC indicators (Eurostat, 2025) rely on self-reported information and capture more subjective aspects of energy poverty, such as the ability to keep the home adequately warm or cool, the physical condition of the dwelling, and difficulties in paying utility bills.

While single indicators are commonly applied in energy poverty research, they have faced criticism for various limitations. Some scholars highlight their insufficient empirical foundation (Fizaine and Kahouli, 2019) or their overly narrow focus on the issue. For example, the 10% indicator is often reproached for its limited scope and strong sensitivity to fluctuations in energy prices (Moore, 2012; Liddell et al., 2012; Legendre and Ricci, 2015). The LIHC measure does not adequately reflect rising energy costs and poses challenges for tracking the effectiveness of policy measures (Heindl and Schuessler, 2015; Marz, 2018). The 2M indicator is criticized for difficulties in objectively defining the minimum income threshold (Castaño-Rosa et al., 2019). Meanwhile, the EU-SILC subjective indicators have been questioned due to their reliance on personal perceptions, cultural influences, and limited applicability across different climatic contexts, which undermines their universality (Bouzarovski, 2014; Castaño-Rosa et al., 2019).

Each single indicator reflects a different conceptual understanding of energy poverty and varies in terms of data requirements, interpretability, and applicability in different socioeconomic and climatic contexts. As no universally accepted measure exists, it is important to understand how individual indicators perform against a set of practical assessment criteria. Table 4.1 therefore contrasts the most widely applied single indicators by examining their strengths and limitations according to ease of calculation and communication, objectivity, data intensity, inclusion of other basic needs, ability to capture hidden deprivation, sensitivity to housing and climatic conditions, suitability for international comparison, and adaptability to other national standards.

Table 4.1 Comparison of the most widely used single indicators for measuring energy poverty. Source: developed by the author.

Indicator	*10%*	*LIHC*	*MIS*	*2M<*	*AFCP*	*EU-SILC*
Easy to calculate	✓	○	○	○	○	○
Easy to communicate	✓	○	○	✓	○	✓
Objective indicator	✓	✓	✓	✓	✓	✗
Requires detailed data	✗	✓	✓	✓	✓	✓
Considers other basic needs	✗	✓	✓	✗	✓	✗
Captures hidden / subjective deprivation	✗	✗	✗	✗	✗	✓
Sensitive to climate and housing conditions	✗	✗	✗	✗	✗	✓
Suitable for international comparison	✗	✗	✗	✗	✗	○
Can be adapted to national standards	○	✓	✓	✓	✓	✗

✓ (yes), ✗ (no), ○ (partially)

The *10 percent indicator* classifies a household as energy poor if it spends more than 10 percent of its income on energy. The main advantages lie in its simplicity—it is very easy to calculate and communicate, requires only basic data (income and energy costs), and is objective. However, it does not consider disposable income or other basic needs, fails to capture subjective deprivation, and is insensitive to climate and housing differences. While it is technically transferable across countries, its fixed threshold may not be meaningful everywhere, which limits its value in international comparison. Although the 10 percent indicator can be applied in any country using the same simple calculation, the meaning of spending more than 10 percent of income on energy differs markedly across contexts. Energy prices, income levels, housing quality, climatic conditions, and typical energy consumption patterns vary significantly between countries. For example, households in colder climates naturally have higher heating needs, while those in warmer climates may need more cooling—so similar spending shares reflect different levels of deprivation. Moreover, in low-income countries, even spending less than 10 percent of income on energy may cause hardship due to limited disposable income, whereas high-income households in developed countries might spend more than 10 percent without experiencing energy poverty. Because of these contextual differences, the fixed 10 percent threshold lacks sensitivity, and direct comparisons of energy poverty rates between countries based on this indicator can be misleading, despite the indicator's methodological transferability.

The *LIHC indicator* classifies households as energy poor if they meet two conditions: spending above the national median on energy and having residual income below the official poverty threshold after energy expenses. This dual-criteria approach improves the measure's ability to reflect the actual financial burden energy costs impose on households by accounting for disposable income, rather than relying solely on energy expenditures or income levels. However, the LIHC's reliance on detailed household microdata, including accurate energy spending and income figures, makes it more data-intensive and complex to calculate than simpler indicators. In terms of international comparison, the LIHC indicator presents significant challenges. Its thresholds—specifically the median energy cost and poverty line—are inherently country-specific, reflecting local economic conditions, energy prices, and social policy standards. This context dependence means that the same absolute values or percentages cannot be universally applied across countries without distorting the measure's meaning. As a result, cross-country comparisons using the LIHC indicator are difficult to standardize, limiting its use in international analyses. Despite these limitations, the LIHC remains highly valuable for domestic policy design and targeting vulnerable populations accurately within a given national context, as it captures nuances of economic hardship linked to energy costs more effectively than many other single-indicator approaches.

The *MIS indicator* identifies households as energy poor when their residual income—income remaining after energy expenses—is below a socially established minimum income standard that reflects an acceptable living standard.

This approach offers a normative perspective by directly linking energy poverty to broader social expectations about minimum living conditions, thus providing a comprehensive measure of economic hardship that accounts for energy costs within the wider context of household needs. However, the MIS indicator's applicability across different countries faces substantial limitations. First, the MIS itself is not a universal benchmark; it must be specifically defined for each country or region, reflecting local social norms, cost of living, and policy frameworks. This country-specific calibration results in a lack of standardization, making it difficult to compare energy poverty levels internationally. Furthermore, defining the MIS inherently involves subjective value judgments regarding what constitutes an acceptable standard of living, which can vary across cultures and over time. These normative decisions complicate efforts to update the MIS consistently or to reproduce it in different contexts, restricting its usefulness for cross-country comparative studies. Despite these challenges, the MIS-based indicator remains valuable for capturing the multidimensional nature of energy poverty within national policy settings, especially where social acceptability and minimum standards are central considerations.

The *2M indicator* classifies households as energy poor if their energy expenditures exceed twice the national mean or median value or if their energy spending is more than double the corresponding share of household income or expenditure. This indicator is relatively straightforward to calculate and communicate, relying primarily on income or expenditure survey data. However, its threshold is inherently arbitrary, lacking a clear empirical or normative basis, which limits its precision in capturing actual energy hardship. From an international comparison perspective, the 2M indicator poses significant challenges. Because the threshold is defined relative to national distributions of energy expenditure or income, it is inherently context-specific and reflects country-specific economic structures, energy prices, and consumption patterns. Consequently, the meaning of "twice the median" in one country can differ substantially from that in another, depending on income inequality, energy market conditions, and household behavior. Moreover, this indicator does not adjust for disposable income or consider other essential household needs, nor is it sensitive to variations in climate or housing conditions, which are critical factors influencing energy requirements and costs across countries. As a result, while the 2M indicator offers a useful relative measure of high energy expenditure within individual countries, its capacity to facilitate meaningful cross-country comparisons remains limited. This limitation underscores the importance of complementing relative indicators with other approaches when conducting international analyses of energy poverty.

The *AFCP indicator* identifies households as energy poor when their income remaining after meeting both essential non-energy expenses and energy costs falls below a defined poverty threshold. This approach captures a household's real disposable income and acknowledges the prioritization of basic needs, thereby aligning closely with deprivation-based understandings of energy poverty. By factoring in essential expenditures beyond energy, the AFCP indicator provides a more nuanced and realistic assessment of a household's financial capacity to afford adequate energy services. However, the complexity of calculating the AFCP indicator poses significant challenges, particularly for international comparison. The method demands detailed and disaggregated data on household income and a wide range of essential expenses, which may not be consistently available across countries. Moreover, the definition of what constitutes "basic needs" can vary considerably between cultural, economic, and social contexts, making threshold setting inherently subjective. This variability complicates efforts to establish standardized or comparable poverty lines internationally. Consequently, while the AFCP indicator offers a conceptually strong and comprehensive measure of energy poverty, its practical application is often constrained by data limitations and the difficulty of harmonizing thresholds across diverse national contexts. This limits its usability for broad international comparisons despite its domestic relevance.

The *EU-SILC* subjective indicators assess energy poverty through self-reported experiences, including difficulties in keeping the home adequately warm, the presence of dampness or leaks, and arrears on utility bills. These indicators are advantageous because they are straightforward to communicate, directly capture hidden and subjective aspects of energy deprivation, and are sensitive to specific housing conditions that affect energy vulnerability. By reflecting the lived experience of households, they provide valuable insights beyond purely financial or consumption-based measures. However, these indicators face some challenges when used for international comparison. Their reliance on self-reported data introduces subjectivity, as perceptions of energy hardship can vary widely based on cultural norms, social expectations, and individual attitudes towards discomfort and deprivation. Additionally, the EU-SILC survey methodology and availability are limited to EU member states. This restricts the applicability of these indicators beyond the EU context.

As it can be seen, international comparisons of single energy poverty indicators face significant methodological and contextual challenges. The diversity of economic conditions, social welfare systems, energy markets, climate zones, housing stock, and cultural perceptions across countries affects how energy poverty manifests and is measured. Indicators developed within one national context often rely on thresholds, reference values, or survey instruments tailored to local circumstances, which may not translate well across borders. Also, it is necessary to highlight that also data availability and quality differ widely across countries, especially between developed and developing regions. Detailed household-level data required for multidimensional or income-residual indicators may be sparse or outdated in low-income countries, while standardized, harmonized data collection is more common in regions such as the European Union. Given these limitations, international comparisons of energy poverty often require cautious interpretation. Researchers and policymakers must consider the conceptual underpinnings, data requirements, and contextual relevance of indicators before drawing cross-country conclusions. Composite or multidimensional indices that incorporate multiple dimensions of energy deprivation, adjusted for local context, may offer improved comparability but still face challenges in harmonization and data collection.

However, indicators such as LIHC, MIS, 2M, and AFCP are inherently adaptable to national standards because their thresholds are defined using country-specific values (e.g., median energy expenditure, poverty lines, or minimum income levels). In contrast, the 10% indicator and EU-SILC subjective measures are less adaptable. A 10% indicator uses a fixed threshold and is therefore not inherently adapted to national standards unless the threshold is modified (e.g., 8 % or 12 %) to reflect country-specific conditions. While EU-SILC subjective indicators rely on harmonized survey questions, which are less adjustable to national standards unless a country modifies the survey instrument or adds tailored questions.

4.2.2 Composite Indicators

As the multidimensional character of energy poverty gains increasing recognition, both researchers and policymakers have shifted their focus toward composite indicators capable of capturing its multidimensional nature. Unlike single indicators, which typically measure only one aspect, such as energy expenditure or self-reported deprivation, composite measures integrate several dimensions, including access to modern energy services, affordability, quality of housing, energy efficiency, and usage patterns. This broader framework allows for a more holistic understanding of energy poverty, particularly across heterogeneous socioeconomic and climatic contexts.

The development of composite indicators is driven by limitations inherent in single indicators measures. These limitations include:

- Insensitivity to household heterogeneity and climatic variations. Single metrics often fail to account for differences in energy needs linked to household composition, regional climates, or building characteristics.
- Poor reflection of lived experience. Solely objective metrics may overlook subjective or experiential dimensions, such as discomfort or energy insecurity.
- Narrow focus on limited energy services. Many traditional indicators concentrate primarily on heating or electricity access, neglecting broader aspects such as clean cooking, lighting, or appliance use.
- Limited relevance for comprehensive policy design. Single-dimensional indicators focus on one aspect of energy poverty, which limits their ability to capture the full complexity of the issue. This narrow focus restricts their usefulness for designing integrated policies that address multiple factors simultaneously.

In contrast, composite indicators combine multiple variables—often integrating both quantitative (e.g., income, energy bills, housing characteristics) and qualitative (e.g., self-assessed thermal comfort) data—into a single evaluative framework. This allows for greater policy relevance through multidimensional diagnostics and improved sensitivity to national and local contexts, enabling adaptation of weighting schemes and thresholds.

Therefore, composite indicators are increasingly recognized as essential tools for capturing the complex interactions between structural determinants and household-level vulnerability, thereby providing a nuanced understanding of energy poverty. By integrating multiple dimensions of deprivation, composite indicators assessing energy poverty contribute to advancing social justice and social equity by highlighting disparities in energy access and affordability across different social groups. Such comprehensive perspective supports evidence-based policymaking that aims to design inclusive and targeted interventions, ensuring that the energy transition equitably benefits all segments of society.

Numerous indicator frameworks and composite indices have been developed to measure energy poverty, reflecting the multifaceted nature of the concept and varying significantly depending on the geographic and economic context. The selection of focus areas and specific indicators often diverges between frameworks designed for developed and developing economies, reflecting different underlying energy challenges and policy priorities.

In the context of developing countries, one of the most prominent and widely applied tools is the Multidimensional Energy Poverty Index (MEPI) developed by Nussbaumer et al. (2012). The MEPI has gained recognition for its comprehensive approach to capturing energy deprivation beyond simple income or expenditure metrics, incorporating multiple dimensions such as access to clean cooking fuels, electricity, and modern energy services. It has been extensively utilized to assess energy poverty across a range of developing regions, including South American countries (Santillan et al., 2020; Pablo et al., 2019), various African nations (Ahmed and Gasparatos, 2020; Ozughalu and Ogwumike, 2019; Israel-Akinbo et al., 2018; Olang et al., 2018), and countries in South and Southeast Asia (Qurat-ul-Ann and Mirza, 2021; Mendoza et al., 2019; Mohanty et al., 2018; Sadath and Acharya, 2017).

The MEPI methodology is grounded in the Oxford Poverty and Human Development Initiative (Alkire and Santos, 2010; Alkire and Foster, 2011), which provides a framework for multidimensional poverty measurement. By focusing on the availability and use of modern energy technologies and services, the MEPI aligns with other indices tailored to the specific needs and circumstances of developing countries, where access to basic energy services remains a critical development challenge.

In contrast, energy poverty indicators and composite indices developed for developed economies often emphasize dimensions such as energy affordability, energy efficiency of buildings, household energy consumption patterns, and subjective experiences of energy deprivation. These differences reflect the contrasting energy landscapes and policy concerns in wealthier nations, including the role of energy markets, infrastructure quality, and environmental considerations.

Unlike the developing world, where composite indices such as MEPI are widely recognized and applied, the measurement of energy poverty in developed countries lacks a single, universally accepted composite indicator. This absence primarily reflects the complex, multifaceted nature of energy poverty in high-income economies, where structural factors, market dynamics, and social policies vary substantially across countries and regions.

In developed economies, energy poverty manifests through diverse pathways including inefficient housing stock, fluctuating energy prices, varied energy supply infrastructures, and heterogeneous social welfare systems. Consequently, researchers and policymakers predominantly adopt context-specific composite indices, which are tailored to reflect the unique socioeconomic conditions, energy consumption patterns, and regulatory environments of their respective countries. This approach ensures the inclusion of relevant aspects—economic, social, technical, behavioral, and subjective—that align with national priorities, cultural specific and data availability.

While single indicators such as LIHC measure remain prevalent due to their simplicity and policy relevance, development of multidimensional frameworks incorporating a range of quantitative and qualitative indicators have gained prominence. These bespoke indices enable nuanced assessments of energy vulnerability and facilitate targeted interventions. However, the methodological heterogeneity and context-dependence inherent in these approaches limit direct comparability of energy poverty levels across developed countries.

Thus, composite energy poverty indicators continue to evolve towards more integrated and adaptable measurement tools, balancing the need for comprehensive analysis with the specificity required to capture the local realities of energy deprivation in developed contexts. The diversity of frameworks and indicators underscores the complexity of energy poverty as a sociotechnical phenomenon and highlights the necessity of context-specific measurement tools to inform effective policy interventions.

Drawing upon an extensive review of scientific literature focused on the development of composite energy poverty indicators, Table 4.2 categorizes commonly applied indicators into key focus areas, reflecting the multifaceted nature of energy deprivation in energy poverty measurements. Organized into thematic groups, the indicators capture diverse dimensions of energy deprivation, including financial constraints, socio-demographic vulnerabilities, housing and infrastructure conditions, environmental and climatic factors, behavioral patterns, subjective experiences, and policy and market influences.

- *Income and energy affordability* indicators capture the core economic dimension of energy poverty by assessing households' financial capacity to meet their energy needs. It reflects the burden of energy costs relative to income and overall economic vulnerability, which are critical to identifying populations at risk of energy deprivation due to limited affordability. Accurate measurement of this dimension helps in understanding how energy expenses constrain household welfare and access to essential energy services.

Table 4.2 Classification of indicators used in energy poverty indices. Source: developed by the author.

Group of indicators	*Examples*
Income and energy affordability	Income; Monetary poverty index; Disposable income ratio; 10% indicator; 2M indicator; Heating expenditure; Energy prices (heating; cooling; electricity); Fuel cost; LPG consumption; Percentage of household expenses to energy; Ratio of required energy expenditure to household income; Share of households spending above threshold on energy
Socio-demographic characteristics	Age of households; Education level; Employment status; Household size; Household type; Social class; Long-term unemployment; Overcrowding rate; Tenants; Rent at market price;
Housing and energy infrastructure	Building age; Number of rooms; State of conservation; Energy efficiency indicator; Energy inefficiency indicator; Performance of energy installations; Access to electricity; Type of fuel; Modern cooking fuel; Heating demand
Environmental and climatic context	Heating Degree Days (HDD); Cooling Degree Days (CDD); Extreme weather vulnerability; Air pollution caused by residential energy use; Indoor pollution; Carbon emissions related to household energy consumption
Multidimensional composite indices	Energy deprivation indices; Indices combining affordability and quality dimensions; MEPI;
Behavioral and consumption patterns	Appliance ownership (TV; washing machine; refrigerator; phone); Use of modern fuels; Residential energy consumption; Electricity consumption; Hot water consumption; Behavioral responses to energy price changes; Participation in demand response or energy-saving programs
Subjective energy deprivation	Perceived thermal comfort; Comfort self-assessment; Arrears on utility bills; Self-reported inability to heat or cool home; 3 EU-SILC subjective indicators for energy poverty measurement
Policy, market, and regulatory factors	Labor Market Policy; Access to credit or energy-specific subsidies; Expenditure on social protection; Energy subsidies; Annual electricity switching rates; Energy market participation barriers; Social rental stock; Energy disconnection and arrears in disadvantaged communities

- *Socio-demographic indicators* provide insight into the social factors that influence energy vulnerability. Variables such as age, employment status, education, household size, and tenure status affect both the energy demand and the resilience of households to energy-related challenges. These factors are essential to contextualize energy poverty within broader social inequalities and to identify groups that are particularly susceptible to energy deprivation.
- *The physical characteristics of housing and energy infrastructure* critically shape energy consumption patterns, efficiency, and access. Indicators in this group reflect the quality, age, and energy performance of buildings, as well as availability and type of energy supply. These factors directly impact the level of energy required to achieve adequate comfort and highlight structural barriers that can exacerbate energy poverty.

- *Environmental and climatic conditions* influence household energy needs and the extent of energy poverty. Such indicators capture variables related to local climate, such as heating and cooling requirements, as well as environmental factors like air pollution. Understanding these contextual elements is vital to account for external pressures on energy demand and associated health risks, thus providing a more comprehensive picture of energy deprivation.
- The usage of *composite indices* integrate multiple dimensions of energy poverty, combining economic, social, infrastructural, and environmental aspects into a single measure. Such holistic approach captures the complex and multifaceted nature of energy deprivation, enabling more nuanced assessments.
- *Household energy consumption behaviors and appliance use* offer important information about energy utilization and coping strategies. This group of indicators reflects how households respond to energy costs and availability, including participation in energy-saving programs or use of modern fuels. Behavioral indicators help to reveal adaptive capacities and inefficiencies, thus enriching the understanding of energy poverty dynamics beyond structural factors.
- *Subjective indicators* capture the lived experience and perceptions of energy poverty, such as comfort levels, inability to maintain adequate heating or cooling, and psychological stress related to energy hardship. These measures are critical for understanding the qualitative dimensions of energy deprivation and for complementing objective data with household self-assessments, which reflect actual well-being impacts.
- *Policy, market, and regulatory* factors encompass institutional and market-related variables that influence energy access and affordability. It includes indicators related to social protection, subsidies, market participation barriers, and regulatory frameworks. Assessing these factors is crucial for identifying systemic obstacles and opportunities within energy systems and for informing policies aimed at reducing energy poverty.

By integrating multiple dimensions of deprivation, composite energy poverty indicators provide a comprehensive framework for understanding the complex interplay between economic constraints, social vulnerabilities, infrastructural deficits, environmental factors, and behavioral patterns. While the individual indicators listed may not explicitly capture disparities among social groups, their combined analysis enables identification of energy poverty patterns across different population segments.

This multidimensional approach is vital for addressing issues of social equity and social justice, as it allows for the detection of which households or communities are disproportionately affected by energy deprivation. By highlighting variations in energy access and affordability across socioeconomic and demographic groups, composite indicators inform policies that can prioritize the needs of the most vulnerable and marginalized populations. Consequently, these tools contribute to the design of equitable energy interventions that seek to reduce systemic inequalities and ensure inclusive access to modern energy services.

In this way, composite indicators serve as essential instruments for evidence-based policymaking, enabling integrated strategies that respond to the diverse and intersecting factors contributing to energy poverty. Their application supports the pursuit of sustainable and socially just energy systems, where the benefits of energy access and the ongoing energy transition are distributed fairly across society.

4.3 Conceptual Intersections Between Energy Poverty and Social Equity

Energy poverty and social equity are intrinsically linked concepts within the energy sector, each shaping and reflecting the other in profound ways. Energy poverty, commonly understood as the inability of households to secure adequate energy services for their basic needs, is not merely a technical or economic issue but fundamentally a manifestation of social inequities embedded in energy systems. Conceptually, energy poverty serves as a prism through which the failures and challenges of achieving social equity in the energy sector become visible and measurable.

Social equity in the energy sector is multidimensional, encompassing the four key dimensions of access, affordability, reliability, and participation. These dimensions provide a comprehensive lens for understanding how energy poverty intersects with broader equity concerns:

- *Access.* Energy poverty often stems from structural barriers that limit equitable access to modern, reliable energy services. These barriers can be geographic, economic, or social, disproportionately affecting marginalized groups and reinforcing spatial and socioeconomic divides.
- *Affordability.* The economic dimension of social equity is central to energy poverty. When energy costs consume a disproportionate share of household income, especially among low-income populations, affordability constraints exacerbate deprivation and social exclusion.
- *Reliability.* Beyond mere access, the consistency and quality of energy services are critical. Energy poverty includes experiences of intermittent supply or inadequate infrastructure, which undermine well-being and social inclusion, particularly in vulnerable communities.
- *Participation.* Social equity also involves procedural justice—ensuring that affected populations have meaningful opportunities to participate in decision-making processes regarding energy policy, planning, and governance. Lack of participation often results in policies that fail to address the specific needs and constraints of energy-poor groups.

Understanding energy poverty through the lens of these four social equity dimensions reveals the multifaceted nature of deprivation and the complexity of challenges faced by affected populations. It highlights that energy poverty is not simply about insufficient energy consumption but involves systemic inequities in how energy services are distributed, governed, and experienced.

From a governance perspective, this intersection demands that energy poverty mitigation policies go beyond aggregate improvements to explicitly target equity outcomes across all four dimensions. Governance frameworks that neglect these dimensions risk reinforcing or even deepening existing disparities, as policies might improve average indicators while failing to reach or benefit the most vulnerable. For example, universal subsidies may increase affordability on average but often fail to reach marginalized groups lacking access or participation opportunities.

Furthermore, energy poverty intersects with other social vulnerabilities such as gender disparities, environmental justice concerns, and energy security challenges. Recognizing these interconnections is essential for developing holistic, equity-centered approaches that address the root causes of deprivation rather than its symptoms alone.

The energy sector itself is both a contributor to and a potential remedy for social inequities manifested in energy poverty. Institutional arrangements, market designs, and infrastructure investments shape who benefits from energy services and who bears the burdens. Hence, integrating social equity considerations into energy sector governance is critical for steering energy systems toward a just transition.

To conclude, it is essential to understand how social inequity and energy poverty serve as drivers within their reciprocal relationship. Social inequity and energy poverty are mutually reinforcing phenomena, each acting as a driver that influences and exacerbates the other. Social inequity—manifested through systemic income disparities, social exclusion, discrimination, and limited access to education and employment—creates structural conditions that increase the risk of energy poverty for marginalized groups. These inequities shape who lacks access to affordable, reliable energy services. Conversely, energy poverty itself serves as a driver of social inequity by constraining individuals' opportunities to improve health, educational attainment, and economic participation. The inability to secure adequate energy perpetuates cycles of deprivation and social exclusion, thereby intensifying existing social inequities. Recognizing this bidirectional dynamic is essential for developing integrated policy responses that address both the root causes and consequences of energy poverty and social inequity in a holistic manner.

Therefore, incorporating social equity considerations into the measurement of energy poverty is essential to capture the multifaceted nature of deprivation and vulnerability. Traditional energy poverty metrics often focus on aggregate indicators such as income thresholds or energy expenditures, which may overlook disparities in access, affordability, reliability, and participation among marginalized groups. By integrating social equity dimensions into measurement frameworks, policymakers and researchers can better identify who is most affected, understand the underlying causes, and design targeted interventions that promote fair and inclusive energy outcomes. Such approach allows to ensure that energy poverty alleviation efforts do not merely improve average conditions but actively reduce inequalities and empower vulnerable populations. A comprehensive framework of indicators including social equity considerations are presented in further section, providing guidance for more nuanced assessment and effective policy monitoring.

4.4 Advancing Energy Poverty Measurement Through Social Equity

Measuring energy poverty requires a multidimensional approach that reflects its complex nature and varied determinants. The assessment depends on contextual factors, including household income, consumption habits, specific energy needs, available technologies, energy prices, climatic conditions, and building energy performance. Developing a robust set of indicators necessitates careful selection to capture the core aspects comprehensively.

Importantly, the framework should integrate the core dimensions of social equity ensuring that measurement reflects not only the material conditions of energy deprivation but also the equitable distribution of energy services, the quality and stability of those services, and the inclusion of affected populations in decision-making processes. This alignment strengthens the measurement of energy poverty within the broader normative principles of distributional, procedural, and recognitional justice, thereby providing a more nuanced and socially responsive understanding.

Two primary methodological approaches are commonly employed in energy poverty measurement: expenditure-based methods, which analyze the relationship between household energy costs and income relative to established thresholds, and consensual-based methods, which incorporate subjective assessments such as the ability to pay bills on time or maintain adequate heating and cooling (Thomson et al., 2017). Both approaches have inherent limitations, and their application should be tailored to the research or policy objectives (Marz, 2018).

When constructing a comprehensive indicator framework, several critical principles must guide the process. First, the indicators should embody the contemporary understanding of energy poverty, encompassing not only economic deprivation but also social and environmental dimensions aligned with sustainable development goals. Second, contextual adaptation is essential to ensure relevance across different socioeconomic and geographic settings. Engaging experts in indicator selection and weighting enhances methodological robustness and contextual fit. Third, the framework must employ transparent, reliable, and scientifically sound methodologies to ensure validity and comparability.

The Bellagio STAMP (Statistical Assessment of Measuring Progress) (Pinter et al., 2012) principles offer a valuable foundation for the development of such an integrated framework. Originally designed for sustainable development monitoring, these principles emphasize balanced, multidimensional indicator sets that include social, economic, and environmental components. Applying Bellagio STAMP principles enables the framework to capture the full spectrum of energy poverty experiences, including economic hardship, social exclusion, and environmental sustainability considerations.

To effectively address energy poverty, it is essential to follow a multidimensional approach that integrates economic, social, and environmental considerations while explicitly incorporating principles of social equity and sustainability. The following framework provides a structured approach to measuring energy poverty, capturing the diverse and interconnected factors that influence households' ability to access and utilize energy services. By linking traditional economic indicators—such as income, energy prices, and expenditure burdens—with social dimensions—including thermal comfort, health impacts, household characteristics, and participation in energy decision-making—and environmental aspects—such as energy efficiency, low-carbon transitions, and climate sensitivity—this framework emphasizes the complex interplay between energy access, societal equity, and sustainable development.

Although energy poverty is a core component of energy sustainability (Kaygusuz, 2012), current measurement practices often overlook sustainability dimensions. Table 4.3 provides a framework that addresses this deficiency by incorporating economic, social, and environmental aspects, thereby aligning energy poverty measurement with a modern, holistic conception of the issue—defined as the inability to secure sufficient quantities of high-quality, reliable, affordable, safe, and sustainable energy.

All identified focus areas within the respective dimensions can be systematically grouped into broader categories of indicators. This grouping not only clarifies the components that define the modern understanding of energy poverty but also emphasizes the pivotal role of social equity, justice, and sustainability in shaping energy policy. By presenting these indicators within a coherent framework, the multidimensional nature of energy poverty is made more visible, underscoring the need for policy approaches that integrate economic, social, and environmental considerations. Such integration ensures that energy policies not only alleviate deprivation but also promote long-term resilience, environmental responsibility, and equitable access to the benefits of the low-carbon transition.

4.4.1 Economic Dimension

The Economic dimension reflects households' financial capacity to meet their energy needs, encompassing income, expenditure, affordability, and access to economic support mechanisms. This dimension is essential for understanding the material constraints that contribute to energy poverty and for designing policies that ensure equitable energy access. By considering economic factors, energy poverty assessments can identify the populations most at risk and

Table 4.3 A comprehensive energy poverty measurement framework integrating social equity and sustainability. Source: developed by the author.

	Focus area	*Examples*
Economic dimension	Income and poverty levels	Income; Monetary poverty index; Disposable income ratio
	Energy prices and affordability	Energy prices (heating price; cooling price; electricity price; domestic hot water price); Fuel cost; LPG consumption
	Energy expenditure share and burden	10% indicator; 2M indicator; Heating expenditure; Percentage of household expenses to energy
	Energy needs and demand	Actual energy needs; Ideal energy needs; Minimum energy expenditure
	Policy effectiveness and regulations	Labor market policy; Access to credit or energy-specific subsidies; Expenditure on social protection
	Energy market dynamics and consumer behavior	Annual electricity switching rates; Final electricity consumption
	Energy infrastructure and fuel type	Access to electricity; Type of fuel; Modern cooking fuel; Energy supply
	Household energy consumption patterns	Residential energy consumption; Electricity consumption; Hot water consumption
	Energy affordability gap	Ratio of required energy expenditure to household income; Share of households spending above threshold on energy
	Macroeconomic drivers	GDP per capita; Inflation rate; Employment rates
	Structural inequalities	Unemployment rates among marginalized groups; Informal sector employment; Income inequality indices; Access to social protection for vulnerable groups; Energy subsidy use by low-income households; Energy disconnection and arrears in disadvantaged communities
	Energy market barriers for vulnerable groups	Limited participation in energy markets; Barriers to switching suppliers; Differential impacts of energy price volatility on low-income or minority households
Social dimension	Subjective deprivation	3 EU-SILC energy poverty indicators
	Thermal comfort	Thermal comfort; Potential heating restriction
	Health and well-being impacts	Health and Quality-of-Life Cost indicator; Indoor pollution
	Demographic and household characteristics	Age of households; Education level; Employment status; Household size; Household type
	Housing tenure and ownership	Ownership of dwelling; Tenure status
	Building characteristics	Year built; Number of rooms; State of conservation
	Energy use habits and lifestyle	Appliance ownership (TV; washing machine; refrigerator); Behavioral responses to energy price changes; Use of energy-saving behaviors
	Geographic location (urban/rural; region)	Location; Region-specific indicators
	Energy literacy and awareness	Public understanding of energy efficiency; Awareness of energy-saving measures

(continued)

Table 4.3 (continued)

	Focus area	*Examples*
	Social cohesion and resilience	Community-energy initiatives
	Energy reliability and quality	Safety; Reliability; Frequency and duration of power outages; Voltage stability; Quality of heating/cooling provision; Consistency of fuel supply
	Temporal dynamics and coping strategies	Seasonal variation in energy consumption and expenditures; Prevalence of coping strategies (rationing, unsafe fuels); Frequency of seasonal energy disconnections
	Participation and agency	Representation of marginalized groups in energy governance; Accessibility of energy-related information; Community engagement levels
	Intersectional equity	Gender-based energy access disparities; Ethnicity-related discrimination in energy services; Age and disability-related barriers in energy use and access
Environmental dimension	Building energy performance and efficiency	Energy efficiency indicator; Energy inefficiency indicator; Performance of energy installations
	Clean energy and low-carbon transition	Renewable energy share; Low-carbon energy consumption structure; Carbon emissions related to household energy consumption
	Environmental impacts of energy use	Air pollution caused by residential energy use; GHG emissions
	Climate sensitivity	Heating degree days; Cooling degree days; Extreme weather vulnerability
	Energy system decarbonization progress	Share of households with access to low-carbon energy; National/regional targets for emissions reduction linked to household energy use
	Access to sustainable energy technologies	Availability and use of energy-efficient appliances; Access to smart meters and energy-management systems; Access to distributed renewable energy systems
	Sustainability beyond emissions	Water consumption in energy production; Resilience of household energy infrastructure to climate change

evaluate the effectiveness of interventions aimed at reducing disparities. These indicators can be arranged into several groups:

- *Household economic capacity and energy needs* (Focus areas: Income and poverty levels; Energy expenditure share and burden; Energy needs and demand; Energy affordability gap). These indicators examine the foundational relationship between household financial resources and the cost of meeting adequate energy needs. They capture the core distributive justice aspect of energy poverty: the unequal capacity to pay for essential services. By considering actual energy expenditure alongside normative benchmarks (minimum, ideal, or required energy needs), both visible and hidden deprivation are identified. From a sustainability perspective, ensuring households can consistently meet their energy needs without financial strain reduces long-term vulnerability and supports stable, inclusive economic development.

- *Energy infrastructure and consumption* (Focus areas: Energy infrastructure and fuel type; Household energy consumption patterns). Access to modern, clean, and reliable energy infrastructure is essential for both equity and sustainability. Many marginalized communities experience infrastructural exclusion, relying on inefficient or polluting fuels. This group connects infrastructure availability to consumption behavior, highlighting the need for targeted investments that enable equitable access to low-carbon, high-efficiency energy systems. Addressing these disparities aligns social justice with climate mitigation goals.
- *Energy market and policy environment* (Focus areas: Energy prices and affordability; Policy effectiveness and regulations; and Energy market dynamics and consumer behavior). Energy markets and policy interventions determine how costs and benefits are distributed across the population. Policy plays a crucial role not only in increasing energy access and affordability but also in shaping household energy efficiency and the development of consumption patterns. It, together with market conditions, influences energy prices, the adoption of technologies, and the feasibility of their implementation (Zhang et al., 2019; Rodriguez-Alvarez et al., 2019; Siksnelyte-Butkiene, 2022). Inefficient, poorly targeted, or regressive policies can exacerbate inequalities, while inclusive, well-designed interventions can promote equity. These indicators emphasize regulatory justice by assessing how market design, tariff structures, and subsidy schemes impact affordability for vulnerable groups. Sustainability here is linked to just transition principles, ensuring that climate and energy policies benefit all socioeconomic levels rather than disproportionately burdening the poor.
- *Broader economic context and inequality* (Focus areas: Macroeconomic drivers; Structural inequalities; Energy market barriers for vulnerable groups). This group of indicators situates energy poverty within the broader economic and social structure. Structural inequalities—rooted in labor market disparities, discrimination, and uneven access to social protection—shape households' exposure to energy poverty. Macroeconomic shocks such as inflation or unemployment compound these risks. From a sustainability perspective, addressing energy poverty without tackling these systemic drivers would risk temporary relief rather than durable change.

4.4.2 *Social Dimension*

The Social dimension captures the non-monetary determinants of energy poverty, reflecting households' living conditions, demographic characteristics, behavioral patterns, and social participation. This dimension is crucial for understanding how energy deprivation intersects with social inequalities and for identifying vulnerable populations. Considering social aspects ensures that energy policies are inclusive, promoting equitable access and participation across different social groups. These indicators can be structured into several categories:

- *Household and individual characteristics* (Focus areas: Demographic and household characteristics; Housing tenure and ownership; Building characteristics). Household composition, tenure status, and dwelling quality significantly influence vulnerability to energy poverty. For example, renters often face restricted access to efficiency upgrades due to split incentives, while older housing stock tends to have lower thermal performance. Linking these structural conditions to social equity ensures that housing policy aligns with energy efficiency goals, creating long-term reductions in both energy poverty and carbon emissions.
- *Energy use behavior and awareness* (Focus areas: Energy use habits and lifestyle; Energy literacy and awareness). This group of indicators captures how households interact with energy through daily practices, habits, and knowledge. It emphasizes the role of behavioral patterns, lifestyle choices, and understanding of energy efficiency in shaping energy consumption. Incorporating social equity, this area highlights disparities in energy literacy and access to information, which can affect households' ability to adopt efficient practices. From a sustainability perspective, promoting informed energy use supports more efficient and low-carbon consumption patterns, contributing to broader environmental objectives.
- *Well-being, comfort, and health* (Focus areas: Subjective deprivation; Thermal comfort; Health and well-being impacts). This group of indicators addresses the human impacts of energy poverty, recognizing that deprivation is not only economic but also physical and psychological. Cold homes, indoor pollution, and thermal discomfort undermine health, productivity, and dignity. Sustainability here is twofold: energy systems should safeguard well-being while minimizing environmental harm, and social policy should address these needs without locking households into high-carbon energy use.
- *Social environment and inclusion* (Focus areas: Social cohesion and resilience; Participation and agency; Intersectional equity). This group of indicators is designed to capture the social dimensions of energy poverty, focusing on the extent to which affected populations are included in decision-making processes, the representation of marginalized groups in energy governance, and the resilience of communities to energy-related shocks and disruptions. Measuring these aspects is essential for understanding not only the material deprivation linked to energy poverty but also the structural and procedural barriers that can perpetuate it. Strong social cohesion can help households cope with energy-related challenges through mutual support networks, while inclusive participation ensures that the specific needs of vulnerable groups are reflected in

energy policies and programs. Intersectional analysis adds further depth, revealing how overlapping vulnerabilities—such as gender, age, ethnicity, disability, and income—shape the experience and persistence of energy poverty. Community-led initiatives, particularly in renewable energy and energy efficiency, provide practical examples of how enhancing inclusion can simultaneously reduce carbon emissions, improve local resilience, and empower disadvantaged populations (Siksnelyte-Butkiene et al., 2023).

- *Energy service reliability, adaptive capacity, and geographic context* (Focus areas: Energy reliability and quality; Temporal dynamics and coping strategies; and Geographic location). This group of indicators captures the quality, stability, and spatial variability of energy services, alongside household responses to service deficiencies. Access alone is insufficient if provision is unreliable, unsafe, or inconsistent; service quality indicators ensure that sustainability goals include resilient, high-quality infrastructure for all, preventing a two-tier energy system. Geographic location shapes these dynamics: rural and remote areas often face higher costs, poorer service quality, and limited infrastructure, highlighting the need for place-based interventions. Simultaneously, households' adaptive behaviors—such as reducing consumption below safe levels or using alternative fuels during seasonal or crisis-driven disruptions—reflect both vulnerabilities and adaptive capacity. Integrating these aspects emphasizes the importance of socially equitable and sustainable energy systems that provide reliable services across all regions while strengthening households' resilience to energy-related challenges.

4.4.3 Environmental Dimension

The Environmental dimension addresses the sustainability and ecological implications of household energy use, emphasizing efficiency, low-carbon transitions, and resilience to climate variability. Integrating environmental aspects allows energy poverty assessments to consider long-term sustainability, ensuring that efforts to improve energy access are compatible with climate and environmental objectives while supporting social equity. To facilitate interpretation, the indicators presented in the framework are arranged in several main categories:

- *Building performance, energy efficiency, and access to sustainable technologies* (Focus areas: Building energy performance and efficiency; Access to sustainable energy technologies). This group of indicators captures the critical intersection of housing quality, technological access, and household vulnerability in shaping energy poverty. Poor building performance, such as inadequate insulation, outdated heating systems, or inefficient design, increases energy costs and carbon emissions, disproportionately impacting low-income households. At the same time, equitable access to energy-efficient appliances, smart meters, and distributed renewable energy sources reduces both financial burdens and environmental impacts. Beyond the technical aspects, these indicators reflect broader issues of social justice: households with inadequate housing or limited technological access face structural disadvantages that hinder participation in the low-carbon transition. Incorporating these dimensions emphasizes interventions that simultaneously improve comfort, reduce energy costs, and enhance environmental sustainability, ensuring that socially vulnerable populations benefit from energy modernization and the clean energy transition.

- *Clean energy transition, decarbonization, and broader sustainability* (Focus areas: Clean energy and low-carbon transition; Energy system decarbonization progress; Sustainability beyond emissions). This group of indicators evaluates clean energy transition, decarbonization, and broader sustainability specifically in the context of energy poverty. It shows the links on how access to renewable or low-carbon energy—particularly in remote or underserved areas—affects vulnerable households by increasing energy availability, reducing dependency on fossil fuels, and mitigating exposure to energy price fluctuations and insecurity. In the scientific literature, renewable energy serves as a critical indicator within the context of energy poverty (Okushima, 2019), affecting many related aspects such as: energy security (Sovacool and Mukherjee, 2011), households' freedom to choose, environmental and air quality, residential comfort, health outcomes (Yadoo and Cruickshank, 2012) among others. Therefore, the expansion of clean energy sources contributes to alleviating both energy poverty and broader social deprivation, supporting the principles of energy justice (Chapman et al., 2018). Moreover, it represents a strategic instrument designed to achieve durable, long-term improvements in energy access, affordability, and environmental sustainability (Terrapon-Pfaff et al., 2015). Indicators in this group also capture environmental aspects of household energy use, including carbon emissions, water consumption in energy production, and resilience of energy infrastructure to climate-related risks. By integrating these measures, these indicators ensure that decarbonization and sustainability efforts contribute directly to alleviating energy poverty, preventing a "green energy divide," and aligning environmental goals with social equity considerations.
- *Environmental and climate impacts* (Focus areas: Environmental impacts of energy use; Climate sensitivity). This group highlights the two-way relationship between energy poverty and environmental stress. Energy-poor households often rely on inefficient or high-emission energy sources, contributing to environmental degradation, while simultaneously being disproportionately vulnerable to climate-related hazards such as heatwaves, cold spells, flooding, or storm damage. Indicators in this cluster assess how environmental risks intersect with household vulnerability, highlighting the need for adaptive strategies that reduce exposure and strengthen resilience. By linking climate sensitivity to energy poverty,

this framework reinforces the importance of interventions that not only minimize environmental harm but also promote equity, ensuring that disadvantaged populations can maintain secure, affordable, and sustainable access to energy in the face of a changing climate.

The developed multidimensional framework provides a comprehensive approach to measuring energy poverty by integrating economic, social, and environmental dimensions while explicitly considering principles of social equity and sustainability. By combining traditional economic indicators—such as income, energy prices, and expenditure burdens—with social measures—such as thermal comfort, health outcomes, household characteristics, and participation in energy decision-making—and environmental indicators—including energy efficiency, clean energy transition, and climate sensitivity—the framework captures the complex interplay of factors that shape households' access to and use of energy. This holistic approach not only enhances the precision and relevance of energy poverty measurement but also ensures that policy interventions can be designed to simultaneously alleviate deprivation, promote social justice, and advance sustainable energy systems. By adopting such an integrated perspective, researchers and policymakers can address the multidimensional challenges of energy poverty, prevent the emergence of new inequalities, and support the transition to energy systems that are equitable, resilient, and environmentally responsible.

References

Abbas, K.; Xie, X.Q.; Xu, D.Y.; Butt, K.M. Assessing an empirical relationship between energy poverty and domestic health issues: A multidimensional approach. Energy, 2021a, 221, 119774.

Abbas, K.; Xu, D.Y.; Li, S.X.; Baz, K. Health implications of household multidimensional energy poverty for women: A structural equation modeling technique. Energy and Buildings, 2021b, 234, 110661.

Ahmed, A; Gasparatos, A. Multi-dimensional energy poverty patterns around industrial crop projects in Ghana: Enhancing the energy poverty alleviation potential of rural development strategies. Energy Policy, 2020, 137, 111123.

Alkire, S.; Foster, J. Counting and multidimensional poverty measurement. Journal of Public Economics, 2011, 95 (7–8), pp. 476–487.

Alkire, S.; Santos, M.E. Acute multidimensional poverty: a new index for developing countries. OPHI Working Paper 38, Oxford Poverty and Human Development Initiative, University of Oxford. 2010, p. 139.

Ambrosio-Albala, P.; Middlemiss, L.; Owen, A.; Hargreaves, T.; Emmel, N.; Gilbertson, J.; Tod, A.; Snell, C.; Mullen, C.; Longhurst, N.; Gillard, R. From rational to relational: How energy poor households engage with the British retail energy market. Energy Research & Social Science, 2020, 70, 101765.

Boardman, B. Fuel Poverty: From Cold Homes to Affordable Warmth. Belhaven Press, London, 1991, p. 267, ISBN 1852931396.

Bouzarovski, S. EP in the European Union: landscapes of vulnerability. Wiley Interdisciplinary Reviews: Energy and Environment, 2014, 3(3), pp. 276–89.

Bouzarovski, S.; Petrova, S. A global perspective on domestic energy deprivation: Overcoming the energy poverty–fuel poverty binary. Energy Research & Social Science, 2015, 10, pp. 31–40.

Castaño-Rosa, R.; Solís-Guzmán, J.; Rubio-Bellido, C.; Marrero, M. Towards a multiple-indicator approach to EP in the European Union: A review, Energy & Buildings, 2019, 193, pp. 36–48.

Chapman, A.J.; McLellan, B. C.; Tezuka, T. Prioritizing mitigation efforts considering co-benefits, equity and energy justice: Fossil fuel to renewable energy transition pathways. Applied Energy, 2018, 219, pp. 187–198.

Charlier, D.; Kahouli, S. From residential energy demand to fuel poverty: income-induced non-linearities in the reactions of households to energy price fluctuations. The Energy Journal, 40 (2), 2019, pp. 101–137.

Dobbins, A.; Nerini, F.F.; Deane, P.; Pye, S. Strengthening the EU response to energy poverty. Nature Energy, 2019, 4(1), pp. 2–5.

Eurostat. European Union Statistics on Income and Living Conditions (EU-SILC). 2025. Available at: [Assessed on 02-06-2025] https://ec.europa.eu/eurostat/web/main/data/database?node_code=livcon

Fizaine, F.; Kahouli, S. On the power of indicators: how the choice of fuel poverty indicator affects the identification of the target population, Applied Economics, 2019, 51(11), pp. 1081–1110.

Garba, I.; Bellingham, R. Energy poverty: Estimating the impact of solid cooking fuels on GDP per capita in developing countries-Case of sub-Saharan Africa. Energy, 2021, 221, 119770.

Groves, C.; Shirani, F.; Pidgeon, N.; Cherry, C.; Thomas, G.; Roberts, E.; Henwood, K. 'The bills are a brick wall': Narratives of energy vulnerability, poverty and adaptation in South Wales. Energy Research & Social Science, 2020, 70, 101777.

Halkos, G.E.; Gkampoura, E.-C. Coping with Energy Poverty: Measurements, Drivers, Impacts, and Solutions. Energies, 2021, 14, 2807.

Heindl, P.; Schuessler, R. Dynamic properties of energy affordability measures. Energy Policy, 2015, 86, pp. 123–132.

Hills, J. Fuel Poverty: The problem and its measurement. Interim Report of the Fuel Poverty Review, 2011, Centre for Analysis and Social Exclusion, Case Report 69, p. 192, ISSN 1465-3001.

Hills, J. Getting the measure of fuel poverty: Final report of the fuel poverty review. Case Report 72, 2012, p. 237, ISSN 1465-3001.

Israel-Akinbo, S.O.; Snowball, J.; Fraser, G. An Investigation of Multidimensional Energy Poverty Among South African Low-Income Households. South African Journal of Economics, 2018, 86(4), pp. 468–487.

Kahouli, S. An economic approach to the study of the relationship between housing hazards and health: The case of residential fuel poverty in France. Energy Economics, 2020, 85, UNSP 104592.

Karakara, A.A., Osabuohien, E.S., Asongu, S. Domestic energy consumption in Ghana: deprivation versus likelihood of access. Management of Environmental Quality: An International Journal, 2021, 32(4), pp. 804–821.

Kaygusuz, K. Energy for sustainable development: A case of developing countries. Renewable and Sustainable Energy Reviews, 2012, 16, pp. 1116–1126.

Kose, T. Energy poverty and health: the Turkish case. Energy Sources Part B-Economics Planning and Policy, 2019, 14(5), pp. 201–213.

Kyprianou, I.; Serghides, D.K.; Varo, A.; Gouveia, J.P.; Kopeva, D.; Murauskaite, L. EP policies and measures in 5 EU countries: A comparative study. Energy and Buildings, 2019, 196, pp. 46–60.

Legendre, B.; Ricci, O. Measuring fuel poverty in France: which households are the most fuel vulnerable? Energy Economics, 2015, 49, pp. 620–628.

Liddell, C.; Morris, C.; McKenzie, S.J.P.; Rae, G. Measuring and monitoring fuel poverty in the UK: National and regional perspectives, Energy Policy, 2012, 49, pp. 27–32.

Liddell, C.; Morris, C.; Thomson, H.; Guiney, C. Excess winter deaths in 30 European countries 1980–2013. A critical review of methods. Journal of Public Health, 2015, 38(4), pp. 806–814.

Lin, B.Q; Wang, Y. Does energy poverty really exist in China? From the perspective of residential electricity consumption. Energy Policy, 2020, 143, 111557.

Marz, S. Assessing the fuel poverty vulnerability of urban neighbourhoods using a spatial multi-criteria decision analysis for the German city of Oberhausen. Renewable and Sustainable Energy Reviews, 2018, 82, pp. 1701–1711.

Mendoza, C.B.; Cayonte, D.D.D.; Leabres, M.S.; Manaligod, L.R.A. Understanding multidimensional energy poverty in the Philippines. Energy Policy, 2019, 133, UNSP 110886.

Mohan, G. Young, poor, and sick: The public health threat of energy poverty for children in Ireland. Energy Research & Social Science, 2021, 71, 101822.

Mohanty, S.K.; Rasul, G.; Mahapatra, B.; Choudhury, D.; Tuladhar, S.; Holmgren, E.V. Multidimensional Poverty in Mountainous Regions: Shan and Chin in Myanmar. Social Indicators Research, 2018, 138(1), pp. 23–44.

Moore, R. Definitions of fuel poverty: implications for policy, Energy Policy, 2012, 49, pp. 19–26.

Nussbaumer, P., Bazilian, M.; Modi, V. Measuring energy poverty: Focusing on what matters. Renewable Sustainable Energy Review, 2012, 16, pp. 231–243.

Okushima, S. Understanding regional energy poverty in Japan: A direct measurement approach. Energy and Buildings, 2019, 193, pp. 174–184.

Olang, T.A.; Esteban, M.; Gasparatos, A. Lighting and cooking fuel choices of households in Kisumu City, Kenya: A multidimensional energy poverty perspective. Energy for Sustainable Development, 2018, 42, pp. 1–13.

Oliveras, L.; Peralta, A.; Palencia, L.; Gotsens, M.; Lopez, M.J.; Artazcoz, L.; Borrell, C.; Mari-Dell'Olmo, M. Energy poverty and health: Trends in the European Union before and during the economic crisis, 2007–2016. Health & Place, 2021, 67, 102294.

Ozughalu, U.M.; Ogwumike, F.O. Extreme Energy Poverty Incidence and Determinants in Nigeria: Multidimensional Approach. Social Indicators Research, 2019, 142(3), pp. 997–1014.

Pablo, Q.S.; Paloma, T.D.; Francisco, J.T. Energy Poverty in Ecuador. Sustainability, 2019, 11, 6320.

Pinter, L.; Hardi, P.; Martinuzzi, A.; Hall, J. Bellagio STAMP: principles for sustainability assessment and measurement. Ecological Indicators, 2012, 17, pp. 20–28.

Primc, K.; Slabe-Erker, R. Social policy or energy policy? Time to reconsider energy poverty policies. Energy for Sustainable Development, 2020, 55, pp. 32–36.

Pye, S., Dobbins, A., Baffert, C., Brajkovic, J., Grgurev, I., De Miglio, R., Deane, P. Energy poverty and vulnerable consumers in the energy sector across the EU: analysis of policies and measures. Policy Report, 2015, p. 77.

Qurat-ul-Ann, A.R.; Mirza, FM. Multidimensional Energy Poverty in Pakistan: Empirical Evidence from Household Level Micro Data. Social Indicators Research, 2021, 155(1), pp. 211–258.

Recalde, M.; Peralta, A.; Oliveras, L.; Tirado-Herrero, S.; Borrell, C.; Palencia, L.; Gotsens, M.; Artazcoz, L.; Mari-Dell'Olmo, M. Structural energy poverty vulnerability and excess winter mortality in the European Union: Exploring the association between structural determinants and health. Energy Policy, 2019, 133, UNSP 110869.

Rodriguez-Alvarez, A.; Orea, L.; Jamasb, T. Fuel poverty and Well-Being: A consumer theory and stochastic frontier approach. Energy Policy, 2019, 131, pp. 22–32.

Sadath, A. C.; Acharya, R. H. Assessing the extent and intensity of energy poverty using Multidimensional Energy Poverty Index: Empirical evidence from households in India. Energy Policy, 2017, 102, pp. 540–548.

Santillan, O.S.; Cedano, K.G.; Martinez, M. Analysis of Energy Poverty in 7 Latin American Countries Using Multidimensional Energy Poverty Index. Energies, 2020, 13(7), 1608.

Siksnelyte-Butkiene, I. Combating Energy Poverty in the Face of the COVID-19 Pandemic and the Global Economic Uncertainty. Energies, 2022, 15, 3649.

Siksnelyte-Butkiene, I.; Streimikiene, D.; Balezentis, T. Addressing sustainability issues in transition to carbon-neutral sustainable society with multi-criteria analysis. Energy, 2022, 254(2), 124218.

Siksnelyte-Butkiene, I.; Streimikiene, D.; Balezentis, T.; Volkov, A. Enablers and barriers for energy prosumption: Conceptual review and an integrated analysis of business models. Sustainable Energy Technologies and Assessments, 2023, 57, 103163.

Siksnelyte-Butkiene, I.; Streimikiene, D.; Lekavicius, V.; Balezentis, T. Energy poverty indicators: A systematic literature review and comprehensive analysis of integrity. Sustainable Cities and Society, 2021, 67, 102756.

Sovacool, B.K.; Mukherjee, I. Conceptualizing and measuring energy security: A synthesized approach. Energy, 2011, 36(8), pp. 5343–5355.

Streimikiene, D.; Lekavicius, V.; Balezentis, T.; Kyriakopoulos, G.L.; Abrham J. Climate Change Mitigation Policies Targeting Households and Addressing Energy Poverty in European Union. Energies, 2020, 13, 3389.

Terrapon-Pfaff, J.; Dienst, C.; Konig, J.; Ortiz, W. How effective are small-scale energy interventions in developing countries? Results from a post-evaluation on project-level. Applied Energy, 2015, 135, pp. 809–814.

Thomson, H.; Bouzarovski, S.; Snell, C. Rethinking the measurement of energy poverty in Europe: a critical analysis of indicators and data. Indoor and Built Environment, 2017, 26(7), pp. 879–901.

Uche-Soria, M.; Rodriguez-Monroy, C. Energy planning and its relationship to energy poverty in decision making. A first approach for the Canary Islands. Energy Policy, 2020, 140, 111423.

United Nations Development Programme. World Energy Assessment Energy and the Challenge of Sustainability, 1st ed.; UNDP: New York, NY, USA, 2000; p. 44.

Yadoo, A.; Cruickshank, H. The role for low carbon electrification technologies in poverty reduction and climate change strategies: A focus on renewable energy mini-grids with case studies in Nepal, Peru and Kenya. Energy Policy, 2012, 42, pp. 591–602.

Zhang, T.; Shi, X.P.; Zhang, D.Y.; Xiao, J.J. Socio-economic development and electricity access in developing economies: A long-run model averaging approach. Energy Policy, 2019, 132, pp. 223–231.

CHAPTER 5

Gender and Energy Equity

5.1 Gender and Energy: Understanding the Nexus

Gender and energy equity is a critical dimension of the global low-carbon transition, yet it remains insufficiently addressed in mainstream energy policy and practice, especially in developing countries. Energy systems are not socially neutral; they are embedded in existing structures of inequality that shape who has access to energy, who participates in decision-making, and who bears the costs of change. Gender, as a system of socially constructed roles, norms, and power relations, influences every stage of the energy chain—from production and distribution to consumption and governance. Without explicit attention to these dynamics, the transition risks reproducing or even deepening existing inequities, undermining both its legitimacy and its long-term sustainability.

5.1.1 Understanding Gender-related Vulnerabilities

The nexus between gender and energy is inherently complex, encompassing household energy consumption patterns, access to renewable and clean energy technologies, sociocultural norms, and the differentiated impacts of energy poverty. At the household level, women often bear primary responsibility for energy management, particularly in cooking and fuel collection. This gendered division of labor has profound implications: reliance on traditional biomass fuels, such as wood or charcoal, not only generates significant health risks—contributing to over three million premature deaths annually, primarily among women and children—but also imposes time and labor burdens that constrain education, income generation, and participation in public life (UN Women and UNIDO, 2023). These patterns illustrate how energy access, or its absence, directly reinforces existing gender inequalities.

Women's limited access to modern energy technologies reflects broader patterns of exclusion from decision-making, employment opportunities, and

I. Siksnelyte-Butkiene, D. Streimikiene, *Social Equity in the Energy Sector*,
https://doi.org/10.1007/978-3-032-22224-4_5

financial resources. Such inequalities manifest across scales: within households (who controls energy-related decisions), within communities (whose needs are prioritized in energy interventions), and within the formal energy sector (where women remain underrepresented in technical and managerial roles). This nexus between gender and energy thus illuminates both the distributional and recognitional dimensions of energy justice.

Access to clean, reliable, and affordable energy presents a critical opportunity to address these inequities. By reducing labor intensity, improving indoor air quality, and enabling the adoption of modern energy technologies, energy interventions can significantly enhance women's health, economic prospects, and overall quality of life. Equally important is the active participation of women in household energy decision-making and broader policy processes. Their perspectives influence energy consumption behaviors, uptake of sustainable technologies, and the formulation of energy policies. Reducing inequities among genders also enables broader social and economic development by enhancing productivity and supporting the expansion of small enterprises. Crucially, these benefits are not automatic. Without deliberate inclusion of women's perspectives in household, community, and policy-level decision-making, energy interventions risk perpetuating existing asymmetries rather than dismantling them.

Access to clean, reliable, and affordable energy can help address these inequities by reducing labor intensity, improving indoor air quality, and enabling adoption of modern technologies. Such interventions can enhance women's health, economic prospects, and quality of life—but these benefits are not automatic. Without deliberate inclusion of women's perspectives in household, community, and policy-level decision-making, energy initiatives risk perpetuating existing asymmetries rather than dismantling them.

Persistent structural barriers—including professional segregation, glass ceilings, and entrenched gender norms—continue to limit women's participation in the formal energy sector (Pueyo and Maestre, 2019; Sunikka-Blank, 2019). Addressing these challenges is essential not only for gender equity but also for the effectiveness and inclusivity of energy transition policies.

Empirical research further highlights gendered patterns in energy behavior and technology adoption across diverse contexts. For instance, a study in Italy revealed that women are more likely than men to engage in energy-efficient behaviors, shaped by local cultural and social norms (Rainisio et al., 2022). In a Chinese province, tiered electricity pricing was found to encourage energy conservation, with women demonstrating higher awareness and adoption of energy-saving practices, particularly when supported by education and family responsibilities (Wu and Zhang, 2017). Similarly, research in Nepal's urban households showed that women were less involved in decisions regarding energy technology purchases, constrained by limited knowledge and financial resources (Shrestha et al., 2021). These findings reinforce the critical role of gender-targeted interventions in promoting both energy efficiency and equity.

A systematic review of 616 publications by Raman et al. (2025) confirms that clean energy adoption can reduce women's unpaid domestic workload, promote leadership opportunities, and enhance economic participation. Women's involvement in energy initiatives also improves household energy efficiency, encourages sustainable energy practices, and increases community engagement. Nevertheless, persistent barriers—such as entrenched gender norms, technology adaptation challenges, and limited financial access—continue to restrict equitable energy outcomes. The study underscores that policies that fail to incorporate gender perspectives risk perpetuating existing inequalities. Integrating gender considerations into energy transition strategies is therefore essential to ensure that technological advancements translate into socially equitable and accessible energy solutions.

The gender–energy nexus has gained prominence in global policy debates, particularly through the interlinkages between United Nations Sustainable Development Goals (SDGs) SDG 5 (gender equality) and SDG 7 (universal access to affordable, reliable, sustainable, and modern energy). Progress in one area is contingent on advances in the other: reducing women's unpaid domestic workload through clean energy access supports gender equality, while women's empowerment is essential for equitable participation in energy system design and governance.

The gender–energy nexus can be understood through several interrelated issues that cut across household, social, and institutional domains. These issues highlight both the disproportionate burdens that women face in contexts of energy poverty and the underutilized opportunities for empowerment that equitable access to energy can unlock (Fig. 5.1).

As illustrated in Fig. 5.1, these issues include household energy responsibilities, adverse health impacts from reliance on biomass fuels, time poverty caused by unpaid domestic work, structural gaps in access to modern energy technologies, exclusion from decision-making processes, persistent inequalities in energy sector employment, and the influence of entrenched sociocultural norms.

At the same time, the unexploited empowerment potential inherent in the energy transition should be underscored: ensuring that women benefit fully from clean and affordable energy not only reduces gender inequities but also strengthens broader societal outcomes, including health, education, productivity, and environmental sustainability. Recognizing these interconnected dimensions is therefore critical for designing energy policies and interventions that are both equitable and effective.

Understanding the nexus between gender and energy therefore requires moving beyond a narrow focus on technological access. It entails recognizing how social equity, cultural norms, and institutional structures shape both the opportunities and constraints that women face in energy systems. It is also necessary to underscore that gender equity is not a "secondary" concern in energy policy but rather a central condition for achieving inclusive and sustainable energy transitions.

Building on this understanding of the gender–energy nexus, the following subsection examines the concept of a gender-responsive, low-carbon transition. It highlights how integrating gender considerations into policies, investments,

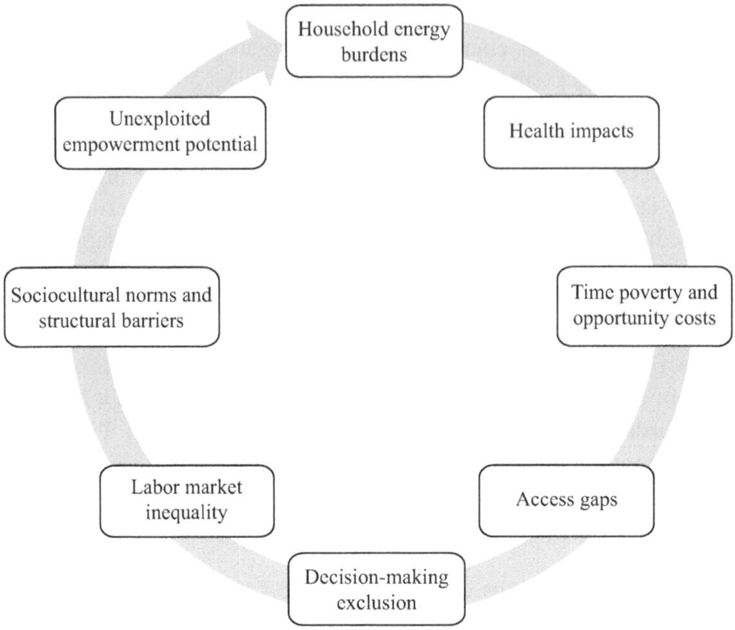

Fig. 5.1 Key issues in the gender–energy nexus. Source: developed by the author.

and governance can help ensure that the shift toward low-carbon, climate-resilient economies can both mitigate disproportionate risks for women and create inclusive opportunities, ensuring that the shift toward climate-resilient, low-carbon economies is socially equitable and transformative.

5.1.2 A Gender-responsive Low-carbon Transition

Gender equity represents a central dimension of social equity in contemporary energy systems. As the global energy sector undergoes rapid transformation—driven by decarbonization targets, technological innovation, and shifts in governance and policy frameworks—opportunities for inclusive participation emerge alongside persistent challenges. Women and marginalized groups continue to experience unequal access to energy resources, limited involvement in decision-making, and constrained control over technologies that shape their daily lives and livelihoods. These disparities are reinforced by intersecting social, economic, and political factors, which can influence who benefits from energy transitions and who remains excluded. Recognizing these dynamics is critical for understanding the broader implications of energy transformations and for informing approaches that seek to align technological advancement with social justice. In this context, exploring gender equity is not merely a normative concern but a practical imperative for building energy systems that are both effective and socially responsive in addressing the needs of diverse communities.

5.1.2.1 *Gendered Impacts of Climate Change and the Just Transition*

Research consistently shows that the consequences of climate change and environmental degradation are not experienced uniformly. Cellini et al. (2025) explored the nexus of gender and energy transitions, identifying seven key dimensions: empowerment, employment/work, attitudes and behaviors, transitions to modern energy, knowledge and awareness, perception, and health. These categories illustrate how gender shapes access to energy, decision-making authority, labor opportunities, social norms, and health outcomes.

Women and girls—particularly those in structurally disadvantaged and marginalized groups such as indigenous women, ethnic minorities, women with disabilities, and women of African descent—face disproportionate risks and impacts (UN Women, 2023; IPCC, 2023). These vulnerabilities are compounded by persistent inequalities in access to decent employment, secure land tenure, natural resources, financial capital, technology, education, mobility, and other enabling assets (Sadiqa et al., 2023). Such systemic barriers limit the capacity to adapt to, mitigate, and recover from climate-related shocks, reducing resilience in the face of environmental crises.

Women are therefore often among those most exposed to the adverse effects of climate change and at heightened risk of exclusion from the processes and benefits of a just transition. Research consistently demonstrates that the consequences of large-scale energy infrastructure projects are not experienced uniformly across genders. Women frequently face disproportionate effects from land-use changes, such as the loss of agricultural plots for small-scale cultivation. Evidence further shows that when projects require involuntary displacement or resettlement, men and women experience distinct economic, social, and environmental risks, with women often disproportionately affected. Even when environmental and social safeguard policies are systematically embedded in international development projects, gender-inclusive public consultations may remain limited, reflecting entrenched social and gender norms within affected communities (Orlando et al., 2018). These patterns underscore the importance of incorporating gender-sensitive planning and governance mechanisms into low-carbon and just transition initiatives to prevent the reinforcement of existing inequalities. At the same time, the low-carbon transition presents an opportunity to redress these disparities: the transformation toward sustainable economies can promote equitable participation and benefits, particularly through the growth of decent work in emerging green sectors—many historically male-dominated.

Realizing this potential requires grounding the just transition in a human rights-based and intersectional framework, with social protection mechanisms at its core. Gender equality and the provision of care must be treated as central pillars of an economy designed to serve both people and the planet. Such an approach demands inclusivity for individuals facing multiple and intersecting forms of discrimination—whether based on sex, income, age, race, ethnicity, gender identity, or sexual orientation—ensuring that no one is left behind.

5.1.2.2 Concept of a Gender-responsive Low-carbon Transition

A gender-responsive low-carbon transition builds on recognition that the global shift toward low-carbon energy systems is not a neutral process. It unfolds within socioeconomic and political systems where gender inequalities remain deeply embedded and can either perpetuate existing disparities or actively dismantle them. Defined here, it refers to the deliberate integration of gender equality considerations into the policies, investments, and governance structures that drive the shift toward low-carbon, climate-resilient economies (UN Women, 2023). It acknowledges that men, women, and gender-diverse individuals experience both climate impacts and decarbonization policies differently due to socially constructed roles, unequal access to resources, and disparities in decision-making power.

Recent global crises—including the COVID-19 pandemic and geopolitical conflicts and wars, as well as climate-caused disasters—have further exposed and deepened gendered vulnerabilities in energy systems (Lazoroska et al., 2024). Rising energy prices, disrupted supply chains, and shifts in labor markets have disproportionately affected women, particularly where barriers to property ownership, finance, and political representation already exist. These dynamics confirm that gender is a structural determinant of who benefits from—and who bears the costs of—the transition.

Ensuring gender equality in low-carbon transitions requires dismantling structural barriers such as informality in women's employment, unequal property rights, and limited access to finance, while creating pathways for women's leadership in emerging green sectors (Anderson and Fisher, 2022). This approach moves beyond simply "adding women" to existing frameworks, aiming instead to transform the systems that perpetuate inequality.

The concept aligns closely with the gender-responsive just transition framework outlined by UN Women (2023), which positions gender equality as both a goal and a condition for achieving sustainable, low-carbon economies. It calls for equal access to decent green jobs, social protection, and skills development, while safeguarding against the disproportionate burdens that poorly designed climate policies can impose on women.

From a theoretical perspective, feminist analyses highlight that gender roles influence career choices, investment capacity, mobility patterns, and energy consumption behaviors, shaping both carbon footprints and the distribution of benefits from decarbonization (Streimikiene, 2023). A gender-just approach therefore addresses root causes of inequality, rebalances power relations, and centers the needs of the most marginalized.

In practice, a gender-responsive low-carbon transition should involve the following:

- Embedding gender analysis into climate and energy policy design;
- Setting measurable targets for women's participation and leadership in green sectors;
- Ensuring equitable access to finance, technology, and training;
- Recognizing and valuing women's knowledge and agency in climate solutions.

By integrating these principles, the low-carbon transition can become not only an environmental imperative but also a driver of social transformation, ensuring that the benefits of decarbonization are shared equitably and that no group is left behind.

5.2 Women's Roles and Challenges in Energy Systems

Women play multiple, often undervalued roles in energy systems, spanning household energy management, community participation, and engagement in small-scale productive activities. They are typically responsible for energy-intensive domestic tasks such as cooking, water heating, fuel collection, and childcare, which require substantial time and labor and may expose them to health and safety risks in certain activities (UN Women and UNIDO, 2023). Despite their critical contributions, women remain underrepresented in formal energy sector employment and leadership positions. Structural barriers—including limited access to education, training, and financial resources, as well as social and cultural norms—further restrict their ability to participate fully in decision-making processes at household, community, and sectoral levels. Understanding these roles and challenges is essential for ensuring social equity in the energy sector, as addressing gender-specific constraints not only empowers women but also promotes inclusive decision-making, equitable access to energy services, and more effective and sustainable energy system outcomes. Figure 5.2 overviews the key constraints faced by women in energy systems.

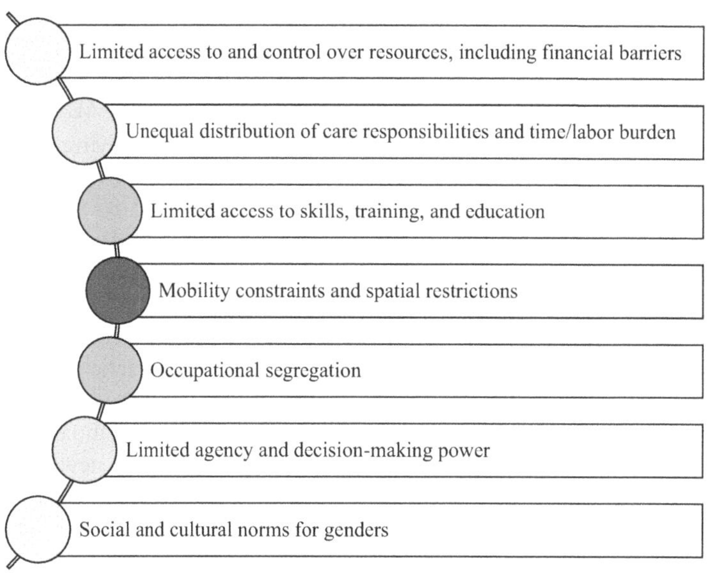

Fig. 5.2 Key constraints for women in energy systems. Source: developed by the author.

5.2.1 *Limited Access to and Control over Resources, Including Financial Barriers*

A fundamental constraint for women in energy systems lies in their limited access to, and control over, productive resources such as land, property, capital, and energy assets, as well as financial services necessary for investment. In many contexts, discriminatory inheritance and property rights constrain women's ability to secure loans, participate in energy-related enterprises, and expand their businesses. Limited access to financial resources further restricts their ability to purchase domestic labor-saving technologies that could free up time for paid work. These structural inequalities are compounded by gender biases within financial institutions, where women entrepreneurs—often concentrated in the informal sector—face greater difficulties in accessing credit, subsidies, or tailored financial products. As a result, women are frequently excluded from investing in modern energy technologies, establishing energy businesses, or scaling up productive uses of electricity. This resource and finance gap not only limits their capacity to benefit directly from energy access but also perpetuates broader patterns of economic marginalization within the sector.

Addressing these barriers requires a combination of legal, financial, and policy interventions. Legal reforms ensuring women's property and inheritance rights, targeted microfinance schemes, and inclusive credit programs can improve access to capital. Moreover, subsidized loans or grants for energy technologies, combined with capacity-building programs, can enable women to invest in productive assets and energy-efficient tools, ultimately enhancing economic participation and empowerment in the energy sector.

5.2.2 *Unequal Distribution of Care Responsibilities and Time/Labor Burdens*

The unequal distribution of unpaid domestic and care work constitutes a structural constraint on women's participation in energy systems. Women bear a disproportionate responsibility for cooking, cleaning, caregiving, and other household tasks. The demanding and time-intensive responsibilities of caregiving, particularly for young children, often force women to select types of work that offer flexibility, allowing them to manage both employment and care duties effectively. These burdens are further intensified in households without modern energy access, where fuel collection, food preparation, and water retrieval require significant time and effort. Time-use studies consistently demonstrate that such responsibilities limit women's ability to participate in education, skill development, and paid employment, while simultaneously exposing them to health risks from household air pollution and heavy manual labor. This "time poverty" reduces women's ability to take advantage of new opportunities created by electrification or renewable energy projects, perpetuating cycles of gendered disadvantage.

Addressing the constraints imposed by unequal care and domestic work responsibilities requires both technological and social interventions. Provision of modern energy services and labor-saving technologies can reduce the time and physical effort required for household energy tasks. Simultaneously, policies and programs aimed at redistributing care responsibilities—such as promoting shared domestic work within households, supporting access to affordable childcare, and encouraging community-based support systems—can enhance women's available time for education, skills training, and income-generating activities. By combining energy access improvements with targeted social measures, it becomes possible to mitigate the gendered impacts of unpaid work, thereby enabling women to engage more fully in energy sector opportunities and decision-making processes.

5.2.3 Limited Access to Skills Development, Training, and Education

Education and skills development represent essential prerequisites for meaningful participation in the energy sector. Studies indicate that women with higher levels of education are less likely to be employed in informal or family-based labor sectors, which typically offer low or subsistence wages and limited job security. Additionally, the quality and accessibility of education significantly influence girls' prospects of obtaining decent employment later in life (Hunt and Samman, 2016). However, women remain significantly underrepresented in science, technology, engineering, and mathematics (STEM) fields, which form the basis for most technical and managerial positions in energy industries. Social norms, gendered stereotypes, and unequal access to education all contribute to this imbalance, resulting in occupational segregation and lower wages for women. Even when women do pursue technical training, barriers in professional networks and mentorship limit their ability to translate skills into career opportunities. Consequently, women are often concentrated in low-skilled or administrative roles, while men dominate decision-making, technical, and leadership positions. Addressing this skills gap is therefore critical to creating inclusive energy transitions.

Addressing these constraints requires multifaceted interventions. Enhancing access to quality education at all levels, promoting girls' participation in STEM disciplines, and providing targeted technical and vocational training are essential. Complementary measures, such as mentorship programs, professional networks for women, scholarships, and policies supporting work-study balance, can facilitate the translation of acquired skills into employment opportunities. Moreover, challenging gendered stereotypes and norms within households, schools, and workplaces can support long-term cultural change, enabling women to access positions of influence and leadership in energy sectors. By addressing both structural and social barriers to education and skills development, energy systems can become more inclusive, harnessing the full potential of women's talents and contributions for equitable and sustainable outcomes.

5.2.4 *Mobility Constraints and Spatial Restrictions*

In many cultural and geographical contexts, women's mobility is limited by a combination of structural and normative factors. Structural barriers include inadequate or unsafe transport infrastructure, poor road conditions, and long distances to schools, energy markets, or training centers. Normative barriers, such as social expectations that women should not travel alone or without male accompaniment, further restrict their movement and access to opportunities. These mobility constraints are closely intertwined with the unequal distribution of household and care responsibilities, where women often dedicate a disproportionate amount of time to domestic chores and caregiving, particularly for children, the elderly, or sick family members, which further limits their ability to participate in education, vocational training, paid employment, or community-based energy initiatives.

The interaction between restricted mobility and domestic obligations reinforces cycles of economic dependence and limits women's capacity to engage fully in energy-related enterprises, sectoral employment, or decision-making processes. Moreover, limited mobility reduces opportunities to attend workshops, access energy services, or participate in collective action, weakening women's voice in shaping energy policies and interventions.

Addressing these constraints therefore requires a multifaceted approach: infrastructural improvements, such as safe and reliable transport; policy interventions ensuring equitable access to training and employment; and social measures that promote a more equitable distribution of household and care responsibilities. By enhancing both mobility and agency, such measures support women's individual empowerment and strengthen their collective role in energy governance and the broader energy sector.

5.2.5 *Occupational Segregation*

Gender-based occupational segregation persists across the energy sector. Women are often concentrated in low-paid, informal, or administrative roles, while technical, managerial, and leadership positions remain dominated by men. This segregation is reinforced by educational disparities, cultural expectations, and institutional biases in hiring and promotion practices.

Global data continues to reveal significant legal barriers to gender equality in the workplace. In 92 economies there is no legal provision requiring equal remuneration for work of equal value. 77 economies impose at least one restriction on the jobs that women can perform at night, in occupations classified as dangerous, or in specific sectors and industries. Within this group, 20 economies prohibit women from working at night, 45 prohibit them from working in jobs deemed hazardous, and 59 still restrict women from working in certain industries—with 18 countries applying such legal restrictions to roles in the energy sector. Examples can be night-work restrictions (e.g., Yemen, Kuwait, Oman, Morocco); hazardous-work restrictions (e.g., Belarus,

Vietnam, Iraq, Russia, Jordan); and energy sector roles restrictions (e.g., Russia, Iran) (World Bank, 2024).

Typical forms of sector-specific legal exclusions in the energy industry can be singled out, which are as follows:

- Underground work bans. These laws prohibit women from working in subterranean environments, often justified on the grounds of safety or health protection. In the energy sector, this can mean exclusion from roles in coal-linked power plants that require underground inspections, hydroelectric tunnels where maintenance crews operate in confined spaces, or underground electrical cabling projects that connect urban grids or industrial facilities. Such bans can prevent women from accessing some of the most technically specialized and well-paid jobs in the industry.
- High-voltage restrictions. Some countries explicitly bar women from working on or near live high-voltage lines or in substation operations where electrical currents exceed certain thresholds. These rules can block women from becoming live-line technicians, grid maintenance engineers, or substation supervisors—positions that are critical for national power reliability and often come with higher wages and career advancement opportunities.
- Oil and gas facility bans. In jurisdictions where power generation is integrated into oil extraction or refining sites, women may be legally excluded from entering these facilities altogether. This can apply to gas-fired power plants located within refinery complexes, offshore platforms with integrated energy systems, or combined heat and power plants serving petrochemical operations. These bans not only limit women's access to the energy sector but also to the lucrative oil and gas industry as a whole.
- Night-shift prohibitions. While not always energy-specific, in countries where the law forbids women from working at night in industrial settings, the impact on the energy sector is significant. Electricity generation and transmission are 24/7 operations, and night-shift crews handle critical maintenance, emergency repairs, and load balancing. Prohibiting women from night work effectively excludes them from half of the operational schedule, reducing their chances for promotion into supervisory or control room roles that require round-the-clock availability.

The consequence of occupational segregation is a persistent wage gap and limited upward mobility for women in the sector. Therefore, addressing these barriers is essential, as sectoral segregation is closely linked to the gender pay gap, and lower salaries for women compared to men can hinder women's opportunities for career advancement (Reshi and Sudha, 2023).

Furthermore, women's underrepresentation in decision-making roles means that their perspectives and needs are frequently excluded from energy project design, implementation, and governance. Without targeted interventions to address occupational segregation, the benefits of energy transitions risk being distributed unevenly, further entrenching gender inequalities.

Campos et al. (2015) found that women entrepreneurs in Uganda operating within male-dominated sectors achieved performance levels comparable to their male counterparts and outperformed women working in female-dominated sectors. The study highlighted the importance of psychosocial factors in enabling women to enter and succeed in male-dominated fields, particularly the presence of male role models and prior exposure to the sector through family or social networks. Despite these successes, women in male-dominated sectors encountered significant challenges, including discrimination from clients and suppliers, threats to business closure, vandalism, and sexual harassment. Broader evidence from the literature suggests that such psychological and social barriers—affecting self-esteem, confidence, and professional recognition—are likely to constrain women across multiple sectors and in diverse national contexts.

Addressing occupational segregation requires multifaceted interventions. Policy measures should promote equitable hiring, career progression, and remuneration practices, alongside targeted support programs such as mentorship, networking opportunities, and leadership training specifically for women. Educational programs that encourage female participation in STEM fields can help reduce long-term structural disparities. Additionally, organizational and sectoral strategies that actively recognize and mitigate gender-based discrimination, bias, and harassment are critical. By addressing these systemic barriers, women can gain access to higher-skilled, better-paid, and decision-making roles, thereby enhancing both individual career trajectories and the inclusivity and effectiveness of energy sector development.

5.2.6 *Limited Agency and Decision-making Power*

Agency and decision-making power represent a crucial dimension of gender equality in energy systems. At the household level, decisions about energy use, technology adoption, and financial investments are often controlled by men, leaving women with limited influence over processes that directly affect their time, health, and economic opportunities. At the community and institutional levels, women remain underrepresented in energy committees, cooperatives, and leadership structures, resulting in policies and projects that may not adequately reflect their needs. An analysis encompassing 72 countries indicated that women occupy merely 6% of ministerial positions responsible for shaping national energy policies and programs (Pearl-Martinez, 2014). Although this study was conducted over a decade ago, such trends still persist, especially in developing countries, reflecting the generally slow pace of change in gender representation within high-level political and policymaking roles. Limited agency also restricts women's ability to negotiate for fair wages, challenge discriminatory practices, or pursue entrepreneurial opportunities. Strengthening women's voice and decision-making power is thus essential for ensuring that energy systems are both equitable and responsive to diverse needs.

Addressing constraints on agency and decision-making power requires interventions at multiple levels. At the household level, awareness-raising campaigns and gender-transformative programs can encourage more balanced decision-making and challenge norms that limit women's participation. At the community and institutional levels, quotas, affirmative action policies, and gender-sensitive governance frameworks can help ensure women's representation in energy committees, cooperatives, and leadership structures. Capacity-building initiatives, including leadership training, negotiation skills, and mentorship programs, are critical to enabling women not only to participate but also to influence decisions meaningfully. Finally, creating inclusive consultation processes in energy planning and project design can ensure that women's needs and perspectives are systematically integrated.

5.2.7 *Gendered Social and Cultural Norms*

Underlying all these constraints are deeply entrenched social and cultural norms that prescribe gender roles and expectations. These norms dictate what kinds of work are deemed appropriate for women and men, shape intra-household decision-making, and influence the design of institutions and policies. For example, cultural beliefs may limit women's involvement in technical fields, restrict their mobility, or reinforce the perception that energy-related decisions are a male domain. Such norms perpetuate inequality across generations, as they affect not only women's current opportunities but also the aspirations and educational pathways of girls. Because social norms interact with economic and structural factors, they are among the most powerful and persistent barriers to gender equality in energy systems.

Addressing the influence of restrictive social and cultural norms requires long-term, multidimensional strategies. Public awareness campaigns, gender-sensitive education programs, and the promotion of positive role models can challenge stereotypes and expand perceptions of what women can achieve. Community-level dialogue and participatory approaches are critical to building local acceptance of women's involvement in nontraditional roles, from technical occupations to leadership positions. At the institutional level, mainstreaming gender considerations into energy policies and programs can help normalize women's participation and ensure their perspectives are included in decision-making. Importantly, interventions must also target men and community leaders, as their support is essential in transforming entrenched norms. By gradually shifting attitudes and expectations, these efforts can create enabling environments that foster women's equal participation in energy systems, thereby strengthening social equity and improving the inclusiveness of energy transitions.

5.2.8 Measures to Address Specific Gender Constraints in Energy Systems

Taken together, the gender-specific constraints in energy systems—ranging from unequal access to resources and financial services, time and care burdens, limited education and skills, mobility restrictions, and occupational segregation to constrained agency and deeply rooted social norms—form a complex web of barriers that systematically limit women's opportunities. These constraints not only undermine women's economic empowerment and well-being but also reduce the efficiency, inclusiveness, and sustainability of energy transitions. Addressing them requires a comprehensive approach that combines infrastructural improvements, targeted capacity-building, institutional reforms, and deliberate efforts to transform discriminatory norms. Ensuring women's full and equal participation in energy systems is not only a matter of social equity but also a prerequisite for leveraging the full spectrum of human capital in designing innovative, just, and sustainable energy solutions. By dismantling these barriers, energy transitions can become more inclusive, equitable, and responsive to the needs of all members of society.

Table 5.1 presents key examples of measures designed to address gender-specific constraints in energy systems. By providing concrete examples within each constraint category, it illustrates how structural, social, cultural, and economic factors are deeply interconnected and how effective strategies must operate across multiple dimensions simultaneously. It also shows the complexity of promoting gender equity in energy systems and the need for coordinated legal, financial, educational, technical, and social measures to create meaningful and sustainable change. Also, by linking specific measures to the challenges identified, it becomes possible to design policies and programs that are both effective and context-sensitive.

Table 5.1 Measures to address specific gender constraints. Source: developed by the author.

Category	*Examples of measures to address gender constraints*
Access to resources and finance	• Legal reforms to ensure women's property and inheritance rights • Targeted microfinance schemes and inclusive credit programs • Subsidized loans or grants for energy technologies • Capacity-building programs for financial literacy and investment in productive assets
Time and care burdens	• Provision of modern energy services and labor-saving technologies • Policies promoting shared domestic work within households • Access to affordable childcare and community support systems • Flexible work arrangements to accommodate caregiving responsibilities
Education and skills development	• Enhance access to quality education at all levels • Promote girls' participation in STEM disciplines • Targeted technical and vocational training programs • Mentorship programs, professional networks, and scholarships • Policies supporting work-study balance and lifelong learning • Challenging gendered stereotypes in society

Table 5.1 (continued)

Category	*Examples of measures to address gender constraints*
Mobility and spatial restrictions	• Infrastructure improvements for safe and reliable transport • Policy interventions ensuring equitable access to training, employment, and energy services • Social measures promoting equitable household and care responsibilities • Support for community-based initiatives and participation
Occupational segregation	• Legal reforms to ensure equitable hiring, promotion, and remuneration • Mentorship, networking opportunities, and leadership training for women • Educational programs encouraging female participation in STEM • Organizational strategies to prevent discrimination, bias, and harassment
Agency and decision-making	• Awareness-raising campaigns and gender-transformative programs • Quotas and affirmative action policies for leadership positions • Gender-sensitive governance frameworks • Leadership, negotiation, and mentorship programs • Inclusive consultation processes in energy planning and project design
Social and cultural norms	• Public awareness campaigns and gender-sensitive education programs • Promotion of positive female role models in technical and leadership roles • Community-level participatory dialogue to challenge stereotypes • Mainstreaming gender considerations in energy policies and programs • Engagement of men and community leaders to support transformative change

5.3 Frameworks and Tools for Gender Mainstreaming in Energy Transition Studies

Gender mainstreaming is widely recognized as a cornerstone of equitable and socially just energy transitions. It involves the systematic integration of gender perspectives into all stages of policy, program, and project cycles, ensuring that women and men—across all their diversity—have equal opportunities to participate in and benefit from the transformation of energy systems. This approach is not merely a matter of fairness; it is a strategic necessity. Without deliberate attention to gender, transitions risk reproducing existing inequalities, such as reinforcing male dominance in technical sectors, overlooking women's specific energy needs, or excluding them from decision-making arenas.

Effective gender mainstreaming requires both conceptual clarity and practical instruments. Conceptual frameworks provide the analytical lenses through which gendered dynamics can be understood, offering a robust basis for identifying inequalities and potential intervention points. Applied tools, in turn, operationalize these insights in concrete policy and project contexts, enabling measurable and effective actions toward gender-responsive energy transitions.

5.3.1 Conceptual Frameworks

A range of conceptual frameworks has been developed to analyze gender dynamics and inform inclusive energy transitions. These frameworks offer insights into how social, economic, and institutional factors intersect with energy systems, enabling more nuanced understanding and strategic planning.

Table 5.2 introduces the most relevant frameworks for gender mainstreaming in energy transition studies.

Table 5.2 Conceptual frameworks for gender mainstreaming in energy transition studies. Created by the author based on: Kimberle (1989); UN Women (2021); Harcourt et al. (2023); Sen (1980); Nussbaum (2000, 2011); Alkire and Deneulin (2009); Robeyns (2017); Kabeer (1994); Overholt et al. (1985); Moser (1993); and UN Women and UNIDO (2023).

Framework	*Purpose*	*Core Elements*	*Relevance to energy transitions*	*Source*
Intersectionality	Analyze overlapping social identities and power structures	Gender, race, class, ethnicity, age, and multiple discrimination	Reveals why certain groups of women face compounded barriers in transition processes	Kimberle (1989); UN Women (2021)
Feminist Political Ecology	Link gender to environmental governance and resource access and power relations	Power relations, environmental knowledge, socioecological systems, access, and control	Highlights gendered impacts of climate and energy transitions, especially in rural or resource-dependent contexts	Harcourt et al. (2023)
Capability Approach	Focus on expanding substantive freedoms and capabilities	Functioning, agency, wellbeing, opportunities	Frames transitions as opportunities to enhance women's real freedoms and quality of life	Sen (1980), Nussbaum (2000, 2011), Alkire and Deneulin (2009); Robeyns (2017)
Social Relations Approach	Analyze how institutions reproduce gender inequality	Rules, resources, activities, people; household, market, state, and community dynamics	Identifies structural and institutional barriers to women's participation in energy decision-making	Kabeer (1994)
Harvard Analytical Framework	Map activities, access, and control by gender	Activity profile, access/control profile, influencing factors	Supports equitable resource allocation in transition projects	Overholt et al. (1985)
Moser Gender Planning Framework	Distinguish practical vs. Strategic gender needs	Triple role (productive, reproductive, community), needs assessment	Ensures transitions address both immediate (e.g., energy access) and structural gender issues (e.g., decision-making)	Moser (1993)

Table 5.2 (continued)

Framework	*Purpose*	*Core Elements*	*Relevance to energy transitions*	*Source*
Gender-Responsive Governance Models	Embed gender equality in decisionmaking and institutional processes	Policy integration, quotas, participatory planning, institutional reform, and monitoring and accountability	Ensures women's voices shape energy transition pathways and governance structures	UN Women and UNIDO (2023)

Intersectionality offers a critical lens for mainstreaming gender perspectives in energy transition studies. Originally articulated by Crenshaw (1989), the concept has been widely applied in academic and policy contexts as a theory, analytical tool, and interpretive framework. It examines how multiple social identities—such as gender, race, class, ethnicity, and age—interact to create overlapping forms of privilege and disadvantage. Rather than viewing these identities in isolation, intersectionality emphasizes their interconnectedness and the ways in which relational dynamics and structural factors produce differentiated outcomes. These intersections are always shaped by prevailing power relations and embedded within broader societal hierarchies (UN Women, 2021). Within the energy sector, applying an intersectional perspective reveals why certain groups of women—particularly those in rural, indigenous, or low-income communities—face compounded barriers in accessing energy services, participating in governance processes, and benefiting from low-carbon technologies. By drawing attention to these layered inequalities, intersectionality strengthens efforts to design energy transitions that are responsive to diverse experiences and more attuned to principles of social equity.

Feminist Political Ecology offers particularly valuable insights for integrating gender perspectives into energy transition studies. The framework links gender to environmental governance and resource access, emphasizing that energy and ecological systems are shaped by unequal power relations. It highlights how women's knowledge, roles, and access to energy-relevant resources are often undervalued, with significant implications for the design and implementation of renewable energy projects and climate policies. Furthermore, the Feminist Political Ecology framework underscores that gendered identities and responsibilities are constructed and negotiated within specific sociocultural and institutional settings, where structural inequalities frequently limit women's agency. In this way, the framework contributes to a deeper understanding of how energy transitions may reproduce or challenge existing social hierarchies. Recent scholarship has recognized its relevance for exploring the intersections of energy, age, and gender (Hanson et al., 2015; Adams et al., 2018; Buechler et al., 2020), reinforcing its value as a lens for addressing questions of equity and inclusion in contemporary energy transitions.

The Capability Approach, developed by Sen (1980) and further elaborated by Nussbaum (2000; 2011), Alkire and Deneulin (2009), and Robeyns (2017), offers a framework for examining development beyond the provision of material resources, focusing instead on individuals' substantive freedoms—the real opportunities that people have to achieve the lives they value. Capabilities can be understood as the substantive freedoms individuals possess to realize their potential in what they can do and be (Robeyns and Byskov, 2020). Applied to energy transitions and gender studies, this approach emphasizes whether women can leverage access to energy to expand their capabilities in multiple domains. For instance, reliable electricity can extend educational opportunities by enabling study after daylight hours; support income-generating activities through small-scale enterprises or home-based work; increase mobility by powering transportation solutions; and facilitate participation in public and political life through access to communication technologies and information networks (Arnaiz et al., 2018). The strength of the Capability Approach lies in its attention to the social, cultural, and structural conditions that determine whether access to energy translates into meaningful empowerment. It foregrounds agency while acknowledging that women's opportunities are unevenly distributed, shaped by intersecting inequalities such as socioeconomic status, ethnicity, age, and geographic location. When combined with intersectional and structural perspectives, the approach allows for a nuanced understanding of how energy transitions can enable or constrain women's real freedoms.

The Social Relations Approach investigates how power is embedded within social relationships and institutions—such as households, markets, and the state—and how these dynamics reproduce gender inequalities. Its focus is not merely on including women in development processes but on transforming the social structures, practices, and relationships that underlie and perpetuate women's disadvantaged positions (Kabeer, 1994). In energy systems, the Social Relations Approach illuminates how formal policies, market interventions, and technological deployments may appear neutral yet reinforce male control over energy resources, decision-making authority, and technological choices. For example, even when electrification projects improve household energy access, entrenched norms regarding who controls energy budgets or who manages appliances may limit women's participation in decision-making. By mapping the distribution of roles, responsibilities, and entitlements, this approach reveals the social dynamics that underpin energy use and access, demonstrating that technological availability alone does not automatically translate into enhanced empowerment or equity. The Harvard Analytical Framework is one of the earliest structured gender analysis tools; it maps differences in activities, access, and control over resources. In energy projects, it identifies who performs energy-related tasks, who controls benefits, and who decides on technology adoption, thereby informing equitable project design.

The Harvard Analytical Framework, developed by Overholt et al. (1985), is one of the earliest structured tools for gender analysis in development projects.

The framework systematically maps differences in activities, access to resources, and control over resources between women and men. Its primary objective is to provide an evidence-based understanding of gender roles, responsibilities, and decision-making power, enabling project planners to design interventions that are more equitable and responsive to the needs of both women and men. In the context of energy projects, the Harvard Analytical Framework identifies who performs energy-related tasks, such as fuel collection, cooking, or maintenance of energy technologies, who controls the associated benefits, and who participates in decisions regarding technology adoption and household energy use. By highlighting disparities in labor, access, and authority, the framework helps ensure that energy interventions do not reinforce existing inequalities and that both women and men can equally benefit from new energy technologies and services. Its structured, data-driven approach has made it a foundational tool for integrating gender considerations into project design and evaluation, providing clear guidance for targeting interventions and measuring outcomes.

The Moser Gender Planning Framework, developed by Caroline Moser (1993), offers a structured approach to integrating gender into development planning by distinguishing between practical and strategic gender needs. Practical gender needs refer to immediate material requirements, such as access to clean cooking fuel, electricity, or efficient stoves, whereas strategic gender needs encompass long-term structural changes, including equitable access to income, land, or decision-making authority. Applied to energy transitions, the framework ensures that interventions address both women's daily necessities and broader empowerment objectives. For example, the provision of efficient cookstoves or household electrification meets practical needs, while facilitating women's participation in energy cooperatives or local governance structures addresses strategic needs, promoting structural transformation and enabling greater agency. The framework highlights the importance of recognizing women's roles and responsibilities in energy use while simultaneously challenging systemic inequalities that limit their participation and control.

Gender-responsive governance models provide a framework for embedding gender equality principles into institutional structures, decision-making processes, and policy implementation. These models emphasize systematic integration of gender considerations to ensure equitable access to resources, participation, and benefits. Within energy transition contexts, gender-responsive governance involves mechanisms such as gender quotas in leadership positions, participatory planning processes that amplify women's voices, and gender-sensitive budgeting to allocate resources equitably. Complementary accountability measures—including gender-disaggregated monitoring and evaluation systems—allow for continuous assessment of policy outcomes and institutional effectiveness. Applying these models in energy governance supports transitions that are not only technically and economically viable but also socially just, inclusive, and responsive to the needs of diverse populations (UN Women, 2021).

5.3.2 *Applied Tools*

A variety of applied tools and planning frameworks have been developed to operationalize gender mainstreaming in energy transitions. These tools translate conceptual insights into concrete methodologies for data collection, analysis, and policy or project design, enabling practitioners to address both immediate energy needs and longer-term structural inequalities. By systematically mapping gendered roles, responsibilities, access to resources, and decision-making power, these tools support inclusive interventions and evidence-based planning.

Applied tools for gender mainstreaming in energy transitions can be grouped by function and application. Capacity-building tools, such as Gender Equality Training (GET), enhance stakeholders' understanding of gender dynamics and equip them to integrate gender perspectives into policies and projects. Impact assessment tools, like Gender Impact Assessment (GIA), systematically evaluate how interventions may differentially affect women and men, guiding program adjustments to promote equity.

Institutional tools, including the Stage-Wise Approach and Gender Audits, support organizations in embedding gender-responsive practices, monitoring progress, and identifying gaps in policies, procedures, and governance structures. Analytical and planning tools, such as the 4R Method (Representation, Resources, Realia, Realization), Gender-Responsive Budgeting (GRB), and Gender Marker Systems, provide structured approaches to assess equity, link financial and operational decisions to gender outcomes, and track project-level gender relevance.

Participatory tools, notably Participatory Gender Appraisal (PGA), engage local communities to identify gendered needs, roles, and priorities, ensuring that decentralized energy interventions are contextually relevant and inclusive.

Table 5.3 presents the most widely used applied tools for integrating gender considerations into energy transition studies, highlighting their practical relevance for policy formulation, project implementation, and monitoring of equitable outcomes.

5.3.2.1 GET

GET is a foundational capacity-building intervention designed to enhance stakeholders' understanding of gender dynamics and to equip them with the necessary competencies to systematically integrate gender perspectives into energy transition policies and projects. By fostering critical reflection on how energy systems intersect with social structures, GET contributes to both individual learning and institutional transformation, positioning gender equality as a central concern in sustainable energy governance.

In practice, GET encompasses a range of pedagogical approaches, including structured awareness-raising workshops, participatory learning exercises, and facilitated reflection sessions that confront implicit and explicit gender biases. Such methods provide stakeholders with the analytical tools to identify how energy interventions may differently affect women and men, while also

Table 5.3 Applied tools for gender mainstreaming in energy transition studies. Created by the author based on: Therivel (2010); Rojas and Siles (2015); CARE (2016); EIGE (2016); OECD (2016); EIGE (2017); Stephenson (2017); EIGE (2019); WFP (2020); and Streimikiene (2023).

Tool	*Purpose*	*Core elements*	*Application in energy transitions*
GET	Strengthen stakeholder capacity to integrate gender perspectives into policy and project design	Awareness raising, skills development, reflection on gender biases	Equips policymakers, project managers, and energy practitioners to design and implement inclusive energy transition strategies
GIA	Systematically assess potential gendered effects of policies, programs, and projects	Relevance screening, data collection and analysis, formulation of recommendations, monitoring	Ensures that energy transition interventions do not exacerbate existing inequalities and informs adjustments for equitable outcomes
Stage-Wise Approach	Track the maturity of institutional gender mainstreaming	Sequential stages: Awareness; Commitment; Capacity; Integration; Sustainability	Monitors organizational progress in embedding gender equality into energy governance structures and project implementation processes
4R Method	Diagnose gender equity across participation, resource access, social norms, and outcomes	Participation, control over resources, lived experiences, and achievement of intended goals	Evaluates whether energy transition benefits are equitably distributed between women and men and informs project adaptation
GRB	Integrate gender considerations into financial planning, allocation, and expenditure	Budget analysis, resource tracking, impact assessment	Ensures that funding for energy projects addresses gender-specific priorities, including women's access to technologies, services, and leadership opportunities
Gender Marker Systems	Classify and monitor the gender relevance of development projects and programs	Standardized scoring system (e.g., 0–2 or 0–3)	Supports funders, policymakers, and program designers in tracking gender integration and prioritizing interventions that promote equality
Gender Audit	Evaluate organizational performance in implementing gender mainstreaming policies and practices	Review of policies and documents, staff interviews, formulation of recommendations	Identifies institutional gaps in energy agencies, utilities, or project teams and informs strategies for capacity-building and organizational reform
PGA	Engage communities in identifying gendered needs, roles, and priorities	Focus groups, participatory mapping, community-based tools	Ensures that local perspectives, particularly of women, inform the design and implementation of decentralized energy projects and transition strategies

offering practical strategies for embedding gender-sensitive perspectives into decision-making processes (EIGE, 2016).

Within energy transitions, GET is particularly significant because it enables stakeholders to anticipate and address barriers to women's participation, access, and empowerment (Rojas and Siles, 2015). For example, training can help planners recognize that women's limited access to land titles or financial credit may constrain their ability to benefit from solar home system programs or that social norms may restrict their participation in energy cooperatives and decision-making bodies. By making these barriers visible, GET supports the design of more inclusive electrification projects, subsidy schemes, and renewable energy initiatives. Furthermore, the institutionalization of GET helps cultivate an organizational culture in which gender equity is treated not as an add-on but as an integral element of effective policy and project planning.

Beyond improving the inclusiveness of energy interventions, GET contributes to broader governance outcomes. It strengthens women's participation in decision-making processes, encourages equitable representation within leadership structures, and fosters accountability in program implementation. By embedding gender-sensitive thinking into the professional practices of energy institutions, GET helps ensure that both practical gender needs (e.g., access to clean cooking technologies or electricity for household enterprises) and strategic needs (e.g., enhanced agency, income opportunities, or leadership roles) are systematically addressed.

The effectiveness of GET, however, depends on several enabling conditions. Sustained institutional support is required to ensure that training is not treated as a one-time exercise but as part of an ongoing process of organizational learning. Reinforcement through follow-up mentoring, refresher courses, or peer-learning networks can help participants translate knowledge into practice. Equally important is the integration of training insights into formal decision-making processes, policies, and performance indicators. Without such mechanisms, the knowledge gained in training sessions risks remaining at the individual level rather than becoming embedded within organizational structures and cultures.

Finally, GET serves as both a technical and transformative tool: it enhances the capacity of energy practitioners to address gender inequalities while simultaneously fostering institutional change that redefines governance practices. In doing so, it contributes to energy transitions that are not only technologically and economically effective but also socially inclusive and equitable.

5.3.2.2 GIA

GIA is a systematic methodology designed to evaluate the potential differential effects of policies, programs, or projects on women and men prior to their implementation. Emerging initially within the field of transport planning, the framework builds on established traditions of Environmental Impact Assessment (EIA), Social Impact Assessment (SIA), and Strategic Environmental Assessment (SEA) but extends these approaches by explicitly integrating considerations of gender equality (Therivel, 2010). This makes GIA both a methodological and normative instrument, as it combines technical assessment with a commitment to equitable outcomes.

At its core, GIA consists of several interrelated components. Relevance screening determines whether gender dimensions are significant to the policy or intervention under review. Data collection and analysis involve the use of sex-disaggregated data and qualitative information to capture the differentiated roles, needs, and constraints of women and men. Formulation of recommendations translates analytical findings into context-specific measures designed to mitigate risks and promote equality. Finally, monitoring mechanisms ensure that gender-related impacts are tracked over time, enabling adaptive learning and corrective actions. Importantly, the approach is strategic and goal-oriented, focusing on how interventions may affect gender equity outcomes and how they can be restructured to foster inclusivity (Streimikiene, 2023).

In the context of energy transitions, GIAs are applied across a wide range of interventions, from national renewable energy strategies and subsidy schemes to decentralized electrification and clean cooking programs. Their utility lies in identifying whether policies and projects risk reinforcing structural inequalities or, conversely, create opportunities for empowerment. For instance, GIAs can reveal whether the deployment of solar home systems primarily benefits male household heads, whether subsidies for clean energy technologies are equitably accessible to women in low-income households, or whether new electrification projects increase women's unpaid labor burdens by shifting responsibilities without altering underlying power relations. By uncovering such differentiated effects, GIAs provide evidence-based insights that guide the adjustment of energy policies and programs to maximize equitable outcomes.

The effectiveness of GIA depends on several enabling conditions. First, data availability remains a central challenge, as many energy initiatives lack sufficiently detailed sex-disaggregated or intersectional datasets. Second, institutional capacity—including staff expertise, resources, and political will—is critical for translating assessment results into actionable recommendations. Third, continuous monitoring throughout the project lifecycle ensures that gender equity objectives are not sidelined once projects move into the implementation phase. Without these conditions, the transformative potential of GIA risks being reduced to a procedural formality.

Ultimately, the value of GIA lies in its ability to bridge the gap between policy intentions and lived realities, ensuring that energy transitions are not only technologically and economically effective but also socially inclusive. By embedding gender-sensitive analysis into the design, implementation, and evaluation of energy initiatives, GIA strengthens the legitimacy, equity, and sustainability of the transition process.

5.3.2.3 The Stage-Wise Approach

The Stage-Wise Approach represents a structured, stage-wise framework for monitoring, guiding, and advancing institutional progress in gender mainstreaming. Although the approach has been adapted across sectors, its core logic remains consistent: gender integration is conceived as a sequential, step-wise process in which each stage builds upon the preceding one, enabling institutions to embed gender-responsive practices sustainably. The principal aim of the approach is to systematize the integration of gender considerations, enhance organizational knowledge and capacity, and provide a coherent road-map for incorporating gender perspectives into policy, governance, and project implementation (Streimikiene, 2023).

Typically, the model includes such stages as: awareness, commitment, capacity, integration, and sustainability. These stages serve as both diagnostic and prescriptive tools. Awareness involves sensitizing staff and stakeholders to gender inequalities and the potential for gender-responsive interventions. Commitment reflects the development of organizational policies and leadership engagement necessary to prioritize gender equality. Capacity encompasses the institutional resources, knowledge, and competencies required to implement gender-sensitive strategies effectively. Integration represents the embedding of gender considerations into formal planning, operational procedures, and decision-making processes. Sustainability ensures that gender mainstreaming practices are maintained over time, institutionalized, and capable of adapting to evolving sociotechnical and policy contexts. There are many different approaches developed by different organizations such as Energia (2011), United Nations Development Programme (2025), and others.

The stage-wise approach allows institutions to break down complex challenges into manageable components, facilitating iterative learning and reflexivity. Organizations can evaluate past achievements, identify gaps in knowledge, resources, or processes, and plan forward-looking interventions that strengthen gender-responsive governance. This iterative reflection is particularly critical in energy transitions, where the sector is characterized by technological complexity, multi-level governance structures, and diverse stakeholder groups.

In energy transition contexts, the Stage-Wise Approach is highly relevant for utilities, energy ministries, regulatory agencies, and project management units seeking to institutionalize gender equality practices. It allows these actors to track progress in integrating gender considerations into strategic planning, program design, and operational procedures. For example, national renewable energy strategies, electrification programs, or energy efficiency initiatives can be evaluated using the model to identify whether awareness-raising has been effective, whether staff possess the necessary capacity, or whether gender considerations are fully integrated into organizational policies.

Moreover, the iterative and reflective nature of the Stage-Wise Approach supports both backward-looking evaluation and forward-looking planning, facilitating continuous adaptation in line with evolving social, technological, and institutional landscapes. By systematically guiding organizations through these stages, the model fosters long-term sustainability of gender mainstreaming, ensuring that gender-responsive thinking is embedded not only in policy and project design but also in the governance culture of energy transition institutions. Ultimately, the Stage-Wise Approach emphasizes that achieving gender equity in energy transitions requires a deliberate, structured, and adaptive approach that accounts for organizational dynamics, sectoral complexities, and the broader socio-political context in which energy systems operate.

5.3.2.4 4R Method

The 4R Method provides a comprehensive assessment of gender equity across four dimensions: representation, resources, realia, and realization. Representation refers to participation in decision-making; Resources considers access to and control over energy-related assets; Realia examines sociocultural and institutional realities shaping gendered experiences; and Realization evaluates the distribution of outcomes and benefits (EIGE, 2019).

In energy transition studies, the 4R Method is particularly valuable for decentralized or community-level interventions such as microgrid development, solar home systems, clean cooking programs, and energy cooperatives. Its multidimensional lens enables practitioners to diagnose not only whether women and men participate equally but also whether they have substantive control over resources and whether interventions produce equitable benefits. For example, a rural microgrid project may provide electricity to households, but without attention to Realia and Resources, women may remain excluded from management committees, training programs, or control over energy usage and financial benefits. Applying the 4R Method helps to reveal such hidden inequities and informs the design of targeted corrective measures, such as capacity-building for women, inclusive governance structures, or policies ensuring equitable allocation of energy resources.

Beyond its diagnostic capacity, the 4R Method serves as a strategic planning tool, linking practical needs with long-term empowerment objectives. By integrating assessments across the four dimensions, energy practitioners can identify leverage points where interventions can simultaneously improve daily energy access and challenge structural barriers to women's participation and decision-making authority. Furthermore, the framework facilitates monitoring and evaluation, allowing implementers to track changes in participation, resource control, and outcomes over time, thereby informing iterative adjustments to interventions.

Critically, the 4R Method emphasizes that energy transitions are not merely technical or economic processes but are deeply socially embedded, shaped by local power relations, gender norms, and institutional practices. It encourages energy planners to consider the interplay between material access and social empowerment, highlighting that technological provision alone does not automatically translate into gender equity. By foregrounding both structural and relational dimensions of inequality, the 4R Method ensures that energy transitions are designed to meet immediate practical needs while fostering strategic empowerment, contributing to socially sustainable and inclusive energy futures.

5.3.2.5 GRB

GRB represents an applied tool for integrating gender considerations into the financial and administrative dimensions of energy policy and project planning. At its core, GRB systematically links budgeting processes—encompassing resource allocation, expenditure monitoring, and financial planning—to gender equality objectives. The approach involves rigorous budget analysis to identify gender biases in allocations, continuous tracking of expenditures to ensure resources reach intended beneficiaries, and assessment of the impacts of financial decisions on gender equity outcomes (EIGE, 2017; Stephenson, 2017).

In the context of energy transition initiatives, GRB serves as a critical mechanism to ensure that investments and financial planning explicitly account for the differentiated needs, opportunities, and constraints of women and men. For example, it can be used to evaluate whether public subsidies for renewable energy technologies, such as solar home systems, micro-hydro installations, or improved cookstoves, are equitably accessible to women, especially those in marginalized or low-income communities. GRB also supports the development of financial instruments that encourage women's participation in energy cooperatives, access to credit for energy entrepreneurship, and inclusion in training and capacity-building programs. By aligning financial flows with equity outcomes, GRB not only enhances the effectiveness of energy interventions but also strengthens the social legitimacy and sustainability of energy transition programs.

Beyond its immediate operational function, GRB plays a strategic role in promoting institutional accountability and transparency. It requires governments, utilities, and project managers to make explicit the gendered impacts of financial decisions, fostering evidence-based planning and enabling stakeholders to identify gaps, inefficiencies, or biases in resource distribution. Successful implementation of GRB is contingent upon the availability of detailed gender-disaggregated financial and programmatic data, institutional commitment to gender equality, and the capacity to adjust allocations or design corrective measures in response to identified inequities.

Moreover, GRB contributes to the broader transformation of energy governance by embedding gender considerations into routine fiscal and administrative practices. It reinforces the principle that achieving gender equity is not merely an aspirational goal but an operational requirement that can be systematically integrated into financial planning and policy execution. In energy transition studies, the tool is particularly relevant for large-scale programs involving public investments, climate finance, or energy infrastructure development, where decisions on resource allocation can either reinforce existing inequalities or create pathways for inclusive economic and social empowerment.

By linking budgetary mechanisms to both practical and strategic gender objectives, GRB ensures that energy transitions not only achieve technological or economic targets but also foster equitable access, participation, and benefit-sharing across diverse social groups.

5.3.2.6 Gender Marker Systems

Gender Marker Systems are standardized classification tools designed to evaluate the extent to which development projects, programs, or policies incorporate gender equality objectives. Originating in the development cooperation context, the most widely applied system is the OECD Gender Equality Policy Marker, which employs a scoring scale (commonly 0–2) to indicate whether gender equality is the principal objective (score 2), a significant but secondary objective (score 1), or not targeted (score 0) (OECD, 2016). Similar marker systems have been adapted by international organizations, multilateral banks, and national agencies to improve transparency, accountability, and comparability of gender integration across sectors and portfolios, for example, the UNDP Gender Equality Seal (2022) and EU Gender Equality Checklists (EIGE, 2024).

In the domain of energy transitions, Gender Marker Systems have increasing relevance for financing renewable energy programs, rural electrification initiatives, energy efficiency measures, and climate–energy nexus projects. They provide funders, policymakers, and implementing agencies with a straightforward instrument to track the degree of gender responsiveness within large-scale funding mechanisms, including climate finance facilities and bilateral energy cooperation schemes. For example, in renewable energy-investment programs, marker systems allow funders to distinguish between projects that explicitly prioritize gender equality (e.g., women-led clean energy enterprises or targeted subsidies for female-headed households) and those where gender is only indirectly addressed. By offering a quantitative benchmark, these systems support evidence-based resource allocation, monitoring, and reporting, thus enabling actors to channel investments toward interventions with greater potential for transformative gender outcomes.

However, while their simplicity and standardization are major strengths, Gender Marker Systems also have limitations. A purely numerical classification risks oversimplifying complex gender dynamics and failing to capture context-specific inequalities. For instance, a project may receive a high score for integrating gender objectives on paper, yet fail to deliver tangible benefits for women in practice due to weak implementation, sociocultural barriers, or lack of community participation. Therefore, effective use of marker systems in energy transitions requires complementing quantitative scoring with qualitative assessments—such as gender audits, participatory gender appraisals, or gender impact assessments—that illuminate local contexts, institutional dynamics, and intersectional differences.

Moreover, the effectiveness of Gender Marker Systems depends on institutional capacity and political will. Accurate application requires trained staff capable of conducting gender-sensitive assessments, robust data systems to document gender-related project outcomes, and governance mechanisms to ensure accountability. Without these enabling conditions, marker systems risk becoming procedural checklists rather than substantive instruments for equity-oriented change.

When applied thoughtfully, Gender Marker Systems can serve as an important entry point for mainstreaming gender into the financing and governance of energy transitions. They enable funders and policymakers to demonstrate commitment to inclusive development, foster alignment with global agendas such as the Sustainable Development Goals (particularly SDG 5 on gender equality and SDG 7 on clean energy).

5.3.2.7 Gender Audit

A Gender Audit is a diagnostic and accountability-oriented methodology designed to evaluate the extent to which organizations and institutions have integrated gender equality into their structures, policies, and practices. The approach is both introspective and participatory, providing institutions with an opportunity to reflect on their internal culture, identify systemic biases, and strengthen their gender mainstreaming efforts. Methodologically, Gender Audits typically combine several techniques: a comprehensive review of policy documents and organizational strategies, structured interviews and focus group discussions with staff at different levels, and the development of tailored, actionable recommendations to close identified gaps.

In the context of energy transitions, Gender Audits are particularly valuable for assessing institutional readiness and capacity within key organizations such as utilities, energy ministries, regulatory bodies, and project management units. For example, a Gender Audit in an energy utility may examine the representation of women in leadership and technical roles, analyze recruitment and promotion practices, assess the gender sensitivity of workplace policies (e.g., parental leave and flexible work arrangements), and review how institutional cultures affect women's participation and career progression.

By systematically assessing organizational structures, processes, and cultures, Gender Audits do more than identify gaps—they also highlight best practices that can serve as models for replication across the energy sector. For instance, an audit might uncover innovative mentoring schemes for women engineers in renewable energy companies or exemplary practices in community engagement by project teams that ensure equitable participation of women and men. Documenting and disseminating such practices strengthens sector-wide learning and facilitates scaling up of gender-responsive approaches.

The application of Gender Audits in energy transitions also has transformative potential at the governance level. By holding institutions accountable for gender mainstreaming, audits contribute to strengthening women's agency and leadership in shaping energy policy and project design. Through targeted recommendations, audits can push institutions to embed gender equality more deeply into their operational frameworks, such as by establishing gender focal points, mandating sex-disaggregated reporting, or integrating gender-sensitive criteria into procurement and project evaluation.

However, the effectiveness of Gender Audits depends critically on two enabling factors: the presence of robust follow-up mechanisms and institutional commitment to act upon recommendations. Without clear accountability pathways and resources allocated to implement corrective measures, audits risk remaining procedural exercises with limited practical impact. Successful audits are those that are embedded in a cycle of continuous improvement, where findings are systematically linked to capacity-building, policy adjustments, and performance monitoring. In energy transition studies, their role extends beyond diagnosis to catalyzing shifts in governance practices, workplace cultures, and institutional accountability mechanisms.

5.3.2.8 PGA

PGA is a community-centered methodological approach designed to actively engage local populations in identifying gendered roles, needs, and priorities. By employing participatory techniques—such as focus group discussions, participatory mapping, household surveys, and community-based analytical exercises—PGA captures the lived realities of women and men in a given context. The approach emphasizes inclusivity, ensuring that marginalized voices, including women, youth, and minority groups, are meaningfully represented in the identification of needs and priorities (CARE, 2016; WFP, 2020).

In the domain of decentralized energy transitions PGAs play a critical role in aligning interventions with local social realities. They provide useful insights into household energy practices, decision-making dynamics, labor burdens, and access to resources, highlighting how energy interventions may differentially affect women and men. For instance, a PGA may reveal that electrification reduces women's domestic labor by decreasing fuel collection time but simultaneously increases unpaid household responsibilities due to expanded expectations for productive activities. Similarly, it can uncover whether women are systematically excluded from energy committee decision-making or whether cultural norms limit their mobility, affecting access to community-based energy services.

By systematically capturing these nuances, PGA findings inform the design of interventions that simultaneously address practical needs—such as access to clean cooking technologies or reliable electricity—and strategic objectives, including enhancing women's agency, participation in governance, and control over resources. Importantly, PGAs support the sustainability of energy projects by ensuring local ownership and legitimacy, fostering community trust, and preempting potential social resistance to new technologies or institutional arrangements.

Effective implementation of PGAs requires careful planning and adequate resources. Skilled facilitation is essential to create safe and inclusive spaces for dialogue, particularly in contexts where traditional hierarchies may limit women's voices. Sufficient time must be allocated to allow iterative engagement, reflection, and validation of findings with the community. Finally, the integration of PGA results into policy formulation, project design, and operational procedures is critical; without institutional commitment to act on community insights, the transformative potential of PGAs is diminished.

Beyond its immediate practical benefits, PGA serves as a strategic tool for enhancing gender-responsive energy governance. By linking local-level social analysis to programmatic and policy decisions, PGAs help bridge the gap between top-down energy planning and community realities.

Each of these methodological tools—ranging from capacity-building interventions such as GET, ex-ante assessments like GIA, stage-wise institutional approaches exemplified by the Stage-Wise Approach, diagnostic frameworks such as the 4R Method, financial instruments like GRB, standardized classification through Gender Marker Systems, introspective institutional reviews in the form of Gender Audits, and community-based participatory processes such as PGAs—provides distinct and context-specific approaches to mainstreaming gender in energy transitions. Depending on project objectives, scale, and institutional capacities, these tools may be applied individually or in selective combination to address different dimensions of governance, from stakeholder learning and organizational accountability to policy design, financial allocation, and community engagement. Their strategic application enhances the identification of gendered dynamics, supports inclusive decision-making, and promotes socially equitable energy interventions. Ultimately, these tools demonstrate that achieving gender-responsive energy transitions requires deliberate, multi-level strategies that embed gender equality as a core principle of policy, practice, and governance, rather than relying solely on technological or economic considerations.

5.4 Empirical Insights on Gendered Energy Dynamics

5.4.1 Key Findings from Gender and Energy Studies

Over the past decade, research on gender and energy has moved beyond theoretical frameworks to incorporate a growing body of empirical evidence. This shift has been driven by the recognition that gendered patterns in energy access, participation, and decision-making are not uniform but instead shaped by intersecting social, economic, and institutional factors. Case studies from diverse geographical contexts, alongside systematic reviews, have provided a more nuanced understanding of how these dynamics unfold in practice.

In their systematic review, which covers 616 studies over a decade (2014–2024), Raman et al. (2025) synthesize the key findings from case studies and categorize them into five key thematic areas, which are presented in Fig. 5.3.

- Reduction of unpaid work. Many studies show that improved access to modern energy services and technologies substantially reduces the time women spend on domestic tasks such as cooking and fuel collection. This time-saving effect is crucial, as it allows women to redirect their efforts toward education, income-generating activities, and community participation.

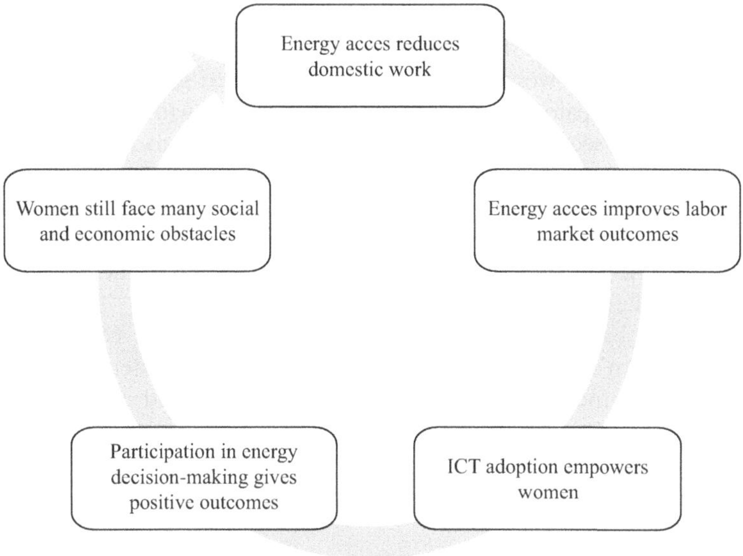

Fig. 5.3 Key findings in the literature addressing gender perspectives in the energy sector development. Source: developed by the author based on Raman et al. (2025).

- Labor market outcomes. Greater access to reliable energy directly influences women's employment and educational opportunities. Electrification and clean energy provision facilitate productive activities, expand opportunities for small enterprises, and support better conditions for balancing household and labor market responsibilities.
- Information and communication (ICT) adoption and empowerment. ICTs play a significant role in enhancing women's agency by increasing access to information, diversifying social networks, and exposing rural households to urban lifestyles, values, and aspirations. In doing so, ICT adoption not only broadens horizons but also supports women in making informed decisions that strengthen their empowerment.
- Decision-making and leadership. Women's active involvement in household and community-energy decisions generates multiple positive effects. Because women often have distinct energy-use priorities and tend to adopt stronger pro-environmental attitudes, their inclusion in decision-making processes improves the uptake of clean and efficient technologies. Nevertheless, despite evidence that women's participation can accelerate the energy transition, they remain underrepresented in managerial, entrepreneurial, and technical roles throughout the energy sector's value chain.

Much research has shown that women's participation in energy entrepreneurship and leadership is often constrained by the perception of entrepreneurship as a male domain. Traditional energy businesses are typically founded and led by men, and even women with significant economic, social, and cultural capital often encounter persistent social and institutional barriers. Intersectional factors further limit opportunities: Indigenous women and other marginalized groups are overrepresented in low-paid, feminized, or precarious roles within energy, mining, and forestry sectors. Barriers such as limited access to financing, lack of local banking infrastructure, and skepticism from financial institutions regarding women-led energy enterprises reinforce these inequalities. Consequently, women are often excluded from entrepreneurship or confined to the least desirable and lowest-paid positions in the sector (Lazoroska et al., 2024).

- Structural barriers. The persistence of socioeconomic constraints that limit women's engagement in energy systems was evident in the literature. These include low purchasing power, restricted access to education and training, and entrenched gender norms that reinforce women's lower social status. Such barriers not only constrain women's agency but also reduce the overall inclusivity and effectiveness of energy transition strategies.

  Women's career opportunities in technical and energy-related fields are heavily shaped by societal perceptions of gender roles. Sociocultural ideas often frame technical occupations as male domains, undermining women's legitimacy and competence, even when their qualifications equal or exceed those of men. Indigenous women and other marginalized groups may face compounded bias due to racialized and gendered stereotypes, leading to tokenization and limited access to desirable positions. Misperceptions about the nature of technical work, such as beliefs that it is socially isolating or overly technical, further discourage women from entering these fields. Workplace cultures may reinforce these barriers through resistance to gender integration, hostile behaviors, and verbal harassment, while "benevolent sexism" can channel women into feminized, lower-status roles rather than offering equal opportunities in demanding positions. Such norms not only limit women's career advancement but also affect work–life satisfaction and their sense of inclusion in professional networks (Lazoroska et al., 2024).

Taken together, these empirical insights reveal that gendered energy dynamics are shaped by a complex interplay of opportunity and constraint. While improved access to modern energy services, ICTs, and inclusive decision-making structures can transform women's economic and social participation, entrenched structural and cultural barriers continue to limit the pace and scope of change. Addressing these challenges requires not only targeted policy interventions and capacity-building initiatives but also a deliberate reshaping of institutional norms and market structures to value women's contributions across the energy value chain. As the energy transition accelerates, ensuring that it is both equitable and inclusive will be critical to unlocking its full social, economic, and environmental potential.

5.4.2 Electrification and Women's Income Generation

Electrification produces distinct gendered effects on labor supply and income generation, reflecting differences in household responsibilities, access to opportunities, and social norms. For women, access to electricity often reduces time-intensive domestic and energy-related tasks. Men's labor patterns, while also influenced by electrification, tend to be less sensitive to these shifts, as their work is often less constrained by domestic energy burdens. Consequently, electricity acts as a gendered technological shock: it reshapes intra-household time allocation, expands productive opportunities for women, and can alter income trajectories, although the exact outcomes remain context-dependent and influenced by local labor markets, social norms, and household decision-making dynamics.

Pueyo and Maestre (2019), in their review of 84 empirical studies, examined the relationship between gender and the productive uses of electricity, with particular emphasis on its implications for labor markets. Their analysis highlighted that electrification generates gender-differentiated effects both on the supply of labor and on the opportunities available in the demand side of labor markets, ultimately shaping income-generation trajectories for women and men in distinct ways. The main gender-specific effects of electrification on women's income generation are identified as:

- positive employment effects for women (increase in employment rate, better work conditions, and enhanced opportunities for home-based enterprises);
- time saving for unpaid work;
- potential impact on earnings and wages;
- shift from agricultural work to non-farm employment after electrification.

From the perspective of Neoclassical Household Economics, electricity functions as a technological shock within the household. Its adoption increases the overall time endowment—through extended lighting and operational hours—while simultaneously enhancing the marginal returns to time spent on domestic work, income-generating activities, and leisure. Consequently, the ultimate patterns of time allocation between men and women emerge from household-level decisions, guided by opportunity costs, individual preferences, and intra-household bargaining dynamics.

Evidence from employment-focused studies consistently indicates that electrification tends to increase women's participation in both formal and informal labor, often more markedly than men's. By reducing time spent on energy-intensive domestic tasks, electricity frees women to engage in paid work and expands opportunities for small-scale, home-based enterprises, enabling additional income generation while accommodating household responsibilities. At the same time, electrification enhances the quality and safety of working conditions, both in households and micro-enterprises, by providing reliable lighting, reducing exposure to harmful fuels, and facilitating the use of modern tools and equipment.

Empirical evidence consistently indicates that electrification facilitates women's transition from agricultural work to non-farm economic activities. This shift is often more pronounced for women than for men, reflecting both the time savings gained from reduced domestic energy burdens and the emergence of new income-generating opportunities. Access to electricity not only enables engagement in small-scale home-based enterprises and service-oriented activities but also supports participation in local labor markets that were previously inaccessible due to energy constraints. By contrast, men's occupational transitions tend to be less responsive to electrification, as their labor is often already concentrated in nonagricultural sectors or less constrained by household energy demands. This gendered differentiation underscores the importance of integrating energy access with broader livelihood and empowerment strategies, ensuring that women can fully capitalize on opportunities created by electrification.

However, the effects of electrification on women's earnings remain context-sensitive. Some case studies report increased income as women take advantage of expanded work opportunities, whereas others note downward pressure on wages due to higher female labor supply in labor markets with constrained demand. These mixed outcomes emphasize that the gendered impacts of electrification cannot be assumed to be universal but must be assessed in relation to local economic structures, labor market dynamics, and prevailing social norms.

Empirical evidence underscores the potential for carefully designed interventions to produce mutually reinforcing benefits: enhancing women's empowerment improves access to clean energy, while improved energy access supports gender equity. Research further indicates that women often demonstrate greater environmental awareness and higher likelihood of adopting sustainable energy practices when empowered to participate in energy-related decision-making.

However, building on the recognition of persistent gendered and social disparities, Johnson et al. (2020) highlight that the deployment of technologies alone does not guarantee inclusive or equitable outcomes in energy transitions. Their findings suggest that no low-carbon energy option inherently produces exclusively positive or negative impacts. Instead, energy technologies are embedded within broader social and political contexts that shape their development, distribution, and utilization. When decision-making processes fail to meaningfully engage rural, low-income, indigenous, or other marginalized communities, these groups often encounter limited agency, reduced ownership, and constrained control over technologies that affect their daily lives and livelihoods. As a result, generic or gender-neutral approaches to community engagement are insufficient to secure equitable outcomes and may inadvertently reinforce preexisting inequalities. These insights underscore the need for energy projects and interventions to actively involve marginalized populations throughout the planning and implementation stages, ensuring that energy transitions advance both technological objectives and broader goals of social justice.

5.4.3 *Gendered Dimensions of Household Energy-saving Behavior*

The intersection of gender and household energy-saving behavior reveals the profound ways in which social roles, responsibilities, and decision-making authority shape energy consumption patterns. Empirical evidence consistently demonstrates that women often undertake a greater share of household energy-related tasks, yet their participation in household energy decision-making remains limited. This asymmetry has important implications for energy efficiency and sustainability outcomes, as the distribution of responsibilities directly influences both energy use and the adoption of conservation practices.

A systematic review by Shrestha et al. (2021) underscores the critical role of gender in shaping energy-saving behaviors. Findings indicate that women generally consume less energy than men in household activities, reflecting differences in task allocation, appliance use, and patterns of energy engagement. Beyond gender, a range of contextual and structural factors—including income, family composition, type and ownership of housing, educational attainment, age, household headship, location, and prevailing sociocultural norms—exerts significant influence on energy-saving practices. These differences are rooted in processes of socialization, normative expectations, and household power relations, collectively shaping choices and behaviors related to energy use.

Key determinants of household energy-saving behavior from a gender perspective include the following (Shrestha et al., 2021):

- Household and socioeconomic context. Building characteristics, lifestyle habits, and socioeconomic conditions substantially shape energy consumption patterns, while cultural norms and values influence women's engagement in energy-related decision-making.
- Demographic and structural variables. Gender, income, type of housing, family composition, location, household headship, age group, ownership, and education all act as determinants of energy-saving behaviors.
- Gendered decision-making processes. Social roles, responsibilities, perceptions, and appliance preferences underpin differences in energy-related choices between men and women.
- Behavioral differences in energy use. Women's generally lower energy consumption in household activities contributes to overall energy efficiency and supports household-level conservation efforts.
- Integrated sustainability strategies. Achieving meaningful energy savings and sustainable outcomes requires a combination of holistic energy planning, gender-focused education, infrastructure development, and systematic mainstreaming of gender considerations within policy and program design.

Analyzing household energy-saving behavior through this gendered lens highlights the importance of policies and interventions that are sensitive to both social and technical dimensions. Gender-responsive strategies not only encourage more efficient energy use but also influence attitudes, habits, and broader participation, fostering both individual and collective engagement in energy conservation.

Promoting equitable participation in energy decisions contributes directly to the achievement of SDG 7 by improving energy security, accessibility, and affordability. Simultaneously, advancing women's agency in household and community-energy matters supports SDG 5, reinforcing gender equality as a core principle of energy transition governance. When gender equity and energy efficiency objectives are pursued together, synergies emerge that can accelerate progress across additional SDGs, reflecting the interconnected nature of sustainable development.

Therefore, household energy-saving behaviors cannot be fully understood or influenced without acknowledging gendered dimensions. Social norms, structural conditions, and behavioral patterns interact to produce differentiated energy outcomes, underscoring the necessity of gender-responsive interventions. Policies and programs that integrate these insights can transform energy use at the household level while simultaneously advancing broader goals of equity, inclusion, and sustainability.

5.4.4 Energy Poverty and Its Multidimensional Impacts on Women

Energy poverty represents a pervasive challenge with significant implications for women's daily lives, limiting their ability to participate effectively in household, community, and economic activities. While energy poverty has been widely studied, research often neglects its gendered dimension, leaving gaps in understanding the specific ways in which women are affected. However, recent research by Pradhan Shrestha et al. (2025) made a significant effort to systematize and categorize the diverse impacts of energy poverty on women.

Pradhan Shrestha et al. (2025) conducted a systematic review of studies examining the consequences of energy poverty for women, revealing multifaceted impacts across several dimensions. The authors categorized these effects into five principal domains: income; health; emotional well-being, increased drudgery, time poverty, and heightened risks of sexual assault; and broader issues of gender inequality and justice.

5.4.4.1 Income Impact

Energy poverty disproportionately affects female-headed households and socially marginalized groups. Women often have lower incomes and limited access to assets yet must allocate a substantial portion of their earnings to energy expenses. This dynamic not only increases their financial vulnerability but also exacerbates broader social inequalities, including reduced access to employment, income-generating opportunities, and decision-making power.

Low-income, restricted ownership of household assets, and limited access to multiple income sources hinder women's ability to adopt modern energy solutions. This dependence generates both direct economic costs and indirect losses: time and labor spent collecting and managing fuel reduce opportunities for income generation and employment. Health consequences of energy poverty further constrain women's labor participation.

Analysis from different contexts (Pradhan Shrestha et al., 2025) revealed a cyclical relationship between energy poverty and unemployment. High energy costs constrain economic opportunities, particularly for women, while income loss limits access to sufficient energy, further entrenching economic vulnerability. Moreover, cultural norms and gendered divisions of labor exacerbate these dynamics, as women's domestic energy responsibilities constrain the time and resources available for formal employment or entrepreneurial activities.

Addressing energy poverty has broader economic implications. Transitioning to modern and affordable energy sources can alleviate women's financial burdens, improve health outcomes, and increase labor participation, thereby enhancing household income and contributing to local and national economic growth. The systemized evidence from different studies (Pradhan Shrestha et al., 2025) underscores that gender-sensitive energy policies are critical not only for reducing household energy vulnerability but also for promoting women's empowerment, social equity, and sustainable economic development.

5.4.4.2 Health Impact

Energy poverty has profound and multidimensional effects on women's health, influenced by both sociocultural norms and the gendered allocation of household responsibilities. In contexts where women are primarily responsible for cooking and domestic energy management, reliance on biomass fuels or other polluting energy sources exposes them to prolonged indoor air pollution, leading to chronic respiratory conditions, chronic obstructive pulmonary disease (COPD), asthma, lung cancer, cataracts, ear diseases, and increased infant mortality. Exposure to biomass smoke also negatively affects maternal and child health, contributing to low birth weight, congenital defects, and reduced productivity.

In addition to respiratory issues, energy poverty contributes to mental health challenges, sleep deprivation, obesity, and overall compromised well-being, particularly when women face combined burdens of domestic drudgery, time poverty, and limited access to modern energy sources. Girls often share these responsibilities from an early age, resulting in educational disruption, restricted opportunities for skill development, and long-term dependence, reinforcing gender inequality.

Regional differences further illustrate the gendered impacts of energy poverty. In the Global South, health consequences are largely linked to indoor air pollution from solid fuels, combined with cold homes in some regions, and amplified by sociocultural norms restricting women's agency. In the Global North, health vulnerabilities are associated with inadequate indoor temperatures, both in winter and summer, disproportionately affecting older women. Energy poverty in this context is linked to respiratory, cardiovascular, and other chronic conditions, with intersectional factors such as age, disability, and socioeconomic status further exacerbating risks.

Cultural attitudes often minimize women's health concerns, and gender-blind energy policies fail to address these disparities, perpetuating energy injustice. Evidence also suggests that the adoption of clean cooking technologies can substantially improve women's health, but uptake depends on affordability, availability, household decision-making dynamics, and social norms (Pradhan Shrestha et al., 2025).

Overall, the literature underscores that energy poverty is not gender-neutral: women bear a disproportionate health burden, both from exposure to harmful fuels and from the sociocultural division of domestic labor (Abbas et al., 2021). Addressing these impacts requires gender-responsive energy policies, clean energy access, and interventions that recognize women's central role in household energy management, linking health outcomes directly to the achievement of SDG 5 and SDG 7.

5.4.4.3 Emotional Well-being Impact

Energy poverty exerts a profound emotional toll on women, largely due to the intersection of gendered household responsibilities and constrained access to adequate energy services. The disproportionate burden of domestic energy tasks exposes women to thermal discomfort, inadequate lighting, and smoke from traditional cooking fuels. This prolonged exposure contributes to heightened levels of depression, anxiety, and psychological distress compared to men (Carrere et al., 2021; Zhang et al., 2022). The chronic nature of energy deprivation fosters a sense of discomfort and alienation within the household, with women often experiencing their own homes as spaces of unease rather than security.

Emotional impacts are further intensified by energy rationing behaviors. Women frequently limit the use of heaters, lights, or cooling devices to prioritize children or other family members, even though research indicates that women are more sensitive to temperature fluctuations than men. Such practices, while protective of family welfare, exacerbate personal discomfort and feelings of guilt and shame. Energy poverty also restricts opportunities for social engagement, as women may avoid hosting guests due to inadequate lighting or temperature control, reinforcing isolation and reducing subjective well-being.

Evidence suggests that energy poverty is associated with broader psychological and social consequences, including chronic stress, financial anxiety, and social marginalization. Various studies indicate a negative relationship between multidimensional energy deprivation and women's life satisfaction and happiness. Women's coping strategies—such as meticulously managing household energy, rationing consumption, or taking on additional domestic responsibilities—demonstrate resilience but impose a substantial mental load and can perpetuate long-term emotional strain.

Therefore, the interventions targeting energy efficiency and access to clean, reliable, and affordable energy are not solely technical or economic in nature. They are crucial for mitigating emotional distress, improving mental health outcomes, and enhancing overall well-being for women (Pradhan Shrestha et al., 2025).

5.4.4.4 *Increased Drudgery, Time Poverty, Heightened Risk of Sexual Assault*

Women disproportionately shoulder the burdens created by energy poverty due to the gendered division of labor in households. It limits their participation in education, employment, and income-generating activities. This reinforces both physical and psychological strains, as inadequate energy access compounds household drudgery and health risks.

In many low-income regions, the collection of firewood is an essential yet highly demanding task assigned largely to women and girls. Such activities consume significant time and contribute to chronic health problems. More alarmingly, firewood collection exposes women to risks of violence and sexual assault (Pradhan Shrestha et al., 2025).

The consequences of these responsibilities extend beyond personal well-being to lost opportunities. Time spent gathering fuel or managing household energy shortages prevents women from pursuing paid work or educational opportunities, reinforcing cycles of poverty.

The findings of the study by Pradhan Shrestha et al. (2025) highlight the need to integrate women into energy policy and decision-making processes: while men often possess more technical knowledge, women tend to prioritize sustainability and social outcomes in energy-related decisions.

5.4.4.5 *Broader Issues of Gender Inequality and Justice*

Much scientific evidence shows that energy poverty is both a driver and a consequence of gender inequality. By limiting women's access to education, health, and income-generating opportunities, it reinforces existing disparities and, in some contexts, gives rise to new forms of exclusion. Cross-country evidence from 51 developing nations demonstrates that alleviating energy poverty strengthens women's socioeconomic and political empowerment, improving their standing in employment, health, and education (Nguyen and Su, 2021). Other global studies similarly confirm that improved access to modern energy enhances women's autonomy and educational achievements.

Beyond its developmental implications, energy poverty is also an issue of justice. Unequal access to energy resources perpetuates systemic inequities, while entrenched cultural norms and household decision-making structures often deny women agency in energy-related choices. This lack of influence deepens their vulnerability, as women bear the costs of energy deprivation without being able to shape solutions.

The manifestations of energy poverty differ across contexts: in the Global South, barriers are largely tied to a lack of electricity and modern cooking fuels, whereas in developed countries affordability and efficiency of energy systems are the predominant concerns. In both cases, however, gender inequality shapes how energy poverty is experienced and sustained.

It is important to note that electrification alone does not resolve these disparities. The transformative potential of energy access depends on how electricity is used—whether for income generation, quality-of-life improvements, or self-development—and whether women are empowered to benefit from it (Pradhan Shrestha et al., 2025).

5.5 Addressing Gender Equity in the Energy Sector

Gender inequalities in energy systems are not accidental but stem from structural, institutional, and sociocultural barriers that systematically disadvantage women. Overcoming gender-specific barriers in energy systems requires a multifaceted set of interventions that combine legal, financial, educational, technological, and cultural strategies. While no single measure can address the complexity of gendered constraints, the coordinated application of reforms and targeted programs can create enabling environments for women's empowerment. Figure 5.4 outlines key groups of measures that have been widely recognized in the literature and practice as crucial for advancing gender equity in energy systems.

5.5.1 Policy and Institutional Measures

Policy and institutional measures form a foundational pillar for promoting gender equity in energy systems. These measures focus on embedding gender considerations directly into governance, planning, and regulatory frameworks, ensuring that women's specific needs, priorities, and barriers are systematically addressed across all stages of the energy policy cycle.

One of the key measures is the mainstreaming of gender in energy laws, policies, and strategies. This involves deliberately incorporating gender perspectives into national and subnational energy frameworks so that policy objectives, program designs, and regulatory standards account for the differentiated impacts of energy decisions on women and men.

Another critical measure is gender-responsive budgeting and governance. This approach allocates financial and institutional resources to initiatives that directly address gender disparities in energy access and employment. By assigning funds for programs targeting women—such as skills training, microfinance for energy enterprises, or clean cooking initiatives—governments and organizations can make tangible progress toward reducing structural inequalities.

Quotas or targets for women's representation in leadership and decision-making bodies address the persistent underrepresentation of women in energy governance. Ensuring women have a seat at the table in regulatory agencies, energy cooperatives, and corporate boards allows them to influence policies, resource allocation, and project priorities. Beyond symbolic representation, these measures enable women to advocate for inclusive decision-making, highlight overlooked needs, and shape energy systems in ways that enhance overall social equity.

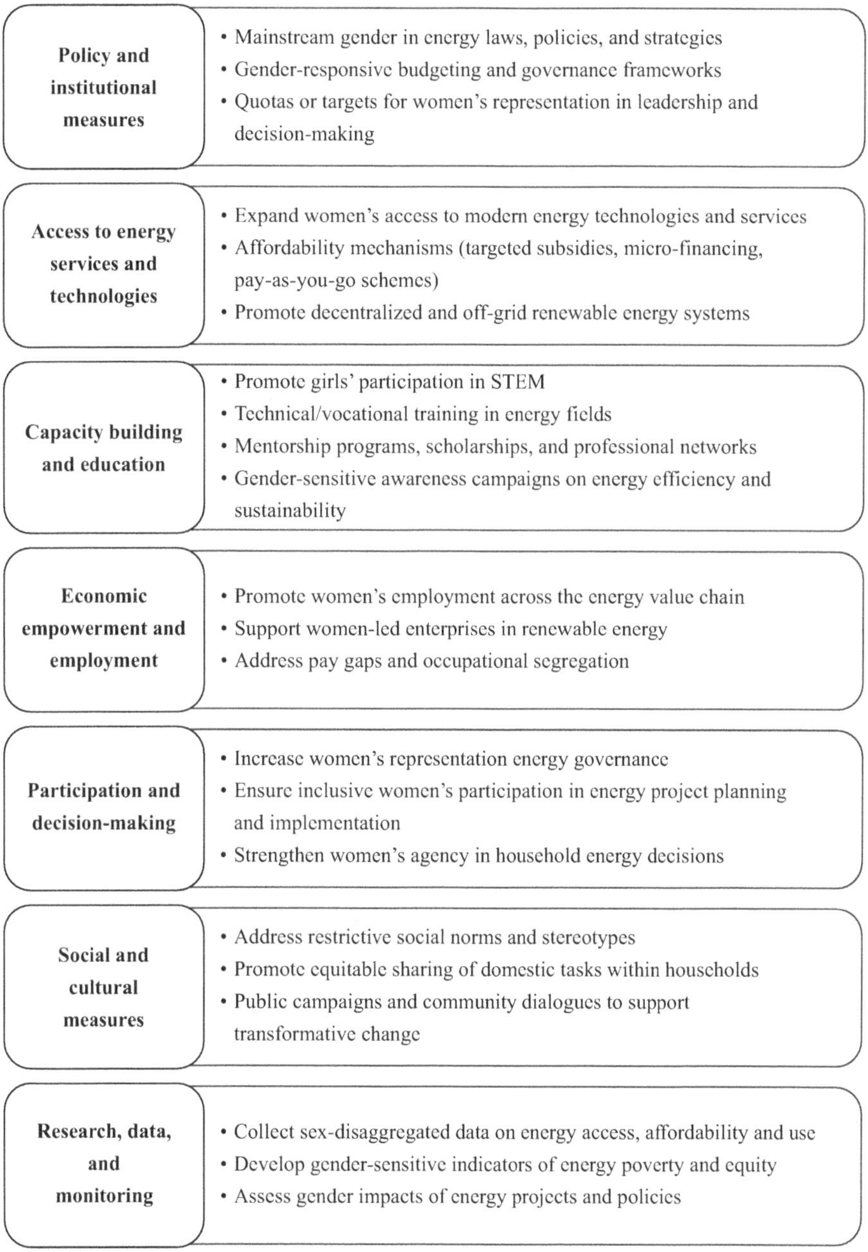

Fig. 5.4 Strategic measures for advancing gender equity in the energy sector. Source: developed by the author.

Policy and institutional measures are strategic because they tackle the root causes of gender inequality in energy systems rather than merely mitigating its effects. By establishing formal structures for women's participation and embedding equity considerations into legal and regulatory frameworks, these measures create an enabling environment for complementary interventions, such as technical training, access to finance, or energy service improvements. While challenges such as resistance to change or weak enforcement exist, evidence shows that countries with stronger gender-responsive policies demonstrate higher female participation in energy decision-making and more equitable outcomes in both energy access and employment.

However, research shows that institutional efforts to address gender disparities in the energy sector still remain fragmented and inconsistent (Lazoroska et al., 2024). Gaps persist both within individual organizations and across energy policy frameworks, often due to limited awareness of the importance of gender considerations and uncertainty regarding how to integrate gender into energy projects and policies. Research highlights the need for proactive, equity-oriented measures, such as employment policies, implementation practices, and organizational reforms that increase gender awareness and promote structural change. Despite rhetoric promoting "meaningful" inclusion, power imbalances often limit who can influence outcomes, reducing the transformative potential of institutional initiatives. Neoliberal economic policies may exacerbate perceptions that gender-focused gains disadvantage other groups, constraining equitable reforms. International examples, such as Brazil's state-funded scholarships for women in science, technology, and engineering, demonstrate that active public sector involvement can expand participation (Baruah and Gaudet, 2022).

5.5.2 Access to Energy Services and Technologies

Access to energy services and technologies represents a critical area for advancing gender equity in energy systems, as it directly affects women's daily lives, productive activities, and economic opportunities. Expanding women's access to modern energy technologies and services—such as electricity, clean cooking fuels, efficient appliances, and water-pumping systems—can substantially reduce the time and labor burdens. By alleviating these constraints, women are better positioned to engage in education, income-generating activities, and community participation, thereby enhancing their overall agency.

Affordability mechanisms are essential for ensuring that women, particularly in low-income households, can benefit from modern energy services. Targeted subsidies, micro-financing options, and pay-as-you-go schemes make clean energy technologies financially accessible and sustainable.

Promoting decentralized and off-grid renewable energy systems constitutes another transformative measure. By bringing energy services closer to communities—through solar home systems, mini-grids, or other renewable solutions—women's mobility constraints and the time spent collecting traditional fuels are significantly reduced. Such systems also allow for flexible energy access that can support both household and entrepreneurial activities, creating new opportunities for income generation and skill development.

Collectively, measures targeting access to energy services and technologies go beyond merely providing energy; they serve as a driver for social and economic empowerment. They address both structural and practical barriers, enabling women to participate more fully in the benefits of the energy transition. Evidence from multiple contexts demonstrates that when women gain reliable, affordable, and locally accessible energy services, their health, productivity, and engagement in decision-making all improve, generating ripple effects that contribute to broader social equity within energy systems.

5.5.3 *Capacity Building and Education*

Capacity building and education are fundamental for promoting gender equity in energy systems, as they empower women with the knowledge, skills, and confidence needed to participate fully in the sector. Encouraging girls' participation in STEM disciplines from an early age addresses the persistent gender gaps in technical fields and lays the foundation for future engagement in energy-related careers. By fostering interest and competence in science, technology, engineering, and mathematics, such measures help counteract social stereotypes that frame energy and technical work as male domains.

Targeted technical and vocational training in energy fields equips women with practical skills required for the installation, maintenance, and management of energy technologies. These programs enhance employability and support entrepreneurship. Mentorship programs, scholarships, and professional networks provide additional avenues for support, enabling women to navigate career pathways, access opportunities, and build confidence in professional settings where they have been historically underrepresented.

Research shows that networks are essential for supporting women's careers and entrepreneurship in the energy sector (Lazoroska et al., 2024). They provide access to knowledge, resources, social support, and legitimacy, helping women overcome barriers to financing, professional advancement, and community recognition. Mentorship complements networks by providing guidance, knowledge transfer, and support for navigating professional hierarchies. Early exposure to networking and mentoring, particularly in STEM-related energy roles, is crucial for reducing gender disparities and fostering long-term professional growth.

Mentors can offer both instrumental support—such as career advice and technical guidance—and psychosocial support, including encouragement, confidence-building, and strategies for overcoming workplace challenges. Mentoring thereby serves as a vehicle for redistributing power and facilitating more equitable access to opportunities.

Gender-sensitive awareness campaigns on energy efficiency and sustainability further complement formal education and training. By raising awareness of women's specific energy needs, promoting the value of their contributions, and disseminating knowledge on sustainable energy practices, such campaigns empower women to make informed decisions at household, community, and institutional levels.

Such educational and capacity-building measures address both structural and attitudinal barriers. They enhance women's human capital, promote equitable participation in energy decision-making, and increase the likelihood that women can leverage the opportunities created by energy transitions, thereby contributing to broader social equity within energy systems.

5.5.4 *Economic Empowerment and Employment*

Economic empowerment and employment measures are crucial to enhance women's agency and participation in the energy sector. Promoting women's employment across the energy value chain—from production and distribution to management and policymaking—helps reduce gender disparities and ensures that women can access opportunities. By deliberately opening pathways into underrepresented roles, these measures challenge occupational segregation and expand women's professional horizons in sectors historically dominated by men.

Supporting women-led enterprises in renewable energy further strengthens economic empowerment by providing women with opportunities to innovate, manage, and scale energy-related businesses. Access to financial instruments, business development services, and capacity-building programs tailored for female entrepreneurs enables them to participate effectively in energy markets, generate income, and contribute to local and national energy solutions.

Research shows that many organizations promote inclusion rhetorically but fail to operationalize it in practice (Wilgosh et al., 2022; Peterson et al., 2022). While gender may be acknowledged as important, this element is frequently treated as a buzzword without corresponding strategies in recruitment, retention, or career advancement. Research indicates that talent identification practices may conflict with stated inclusiveness goals, resulting in less diverse talent pools. Ambiguities in advancement processes and reliance on informal or "hidden" networks often create advantages for men, who typically have greater access to and capacity to leverage such networks, while women encounter barriers to recognition and promotion. Gender discrimination and the unequal influence of informal networks thus reinforce workplace inequities, highlighting the persistent gap between organizational rhetoric and lived workplace realities (Lazoroska et al., 2024). Therefore, addressing pay gaps and occupational segregation is equally critical for fostering equity. Ensuring equitable remuneration, transparent promotion practices, and inclusive hiring policies not only increases women's economic security but also signals the sector's commitment to fairness and diversity.

Together, these measures help create enabling environments where women can translate education, training, and entrepreneurial initiatives into tangible economic outcomes.

5.5.5 *Participation and Decision-making*

A gender-responsive approach to just transition requires inclusive social dialogue and the active participation of all stakeholders concerned. Ensuring women's leadership and substantive involvement in the formulation, implementation, monitoring, and evaluation of just transition policies and programs is essential. Such engagement not only enhances the overall effectiveness of transition initiatives but also safeguards equal opportunities and equitable outcomes for women and girls in all their diversity (UN Women, 2023).

Participation and decision-making measures aim to enhance women's voice and influence across all levels of the energy sector. Increasing women's representation in energy governance structures, such as cooperatives, regulatory bodies, and community-energy committees, ensures that their perspectives and needs are considered in the design, planning, and implementation of energy projects. This representation is essential for creating policies and programs that are responsive to gender-specific challenges and opportunities.

Ensuring inclusive participation in energy project planning and implementation allows women to contribute practical insights, particularly regarding household energy use, productive energy applications, and local energy needs. When women are actively engaged in decision-making processes, projects are more likely to achieve equitable outcomes and foster broader community support.

Strengthening women's agency in household energy decisions complements these institutional measures by empowering women to influence choices about energy technologies, spending, and management within their homes. This can improve the adoption of energy-efficient solutions, support better time allocation, and enhance household welfare.

5.5.6 *Social and Cultural Measures*

Social and cultural measures focus on challenging deeply entrenched norms and stereotypes that limit women's roles in energy systems. Restrictive social expectations often dictate which types of work are considered appropriate for women and men, influence household decision-making, and shape participation in technical or leadership roles. Addressing these norms is essential to remove barriers that prevent women from fully engaging in the energy sector.

One key approach is promoting the equitable sharing of domestic and caregiving responsibilities within households. Policies and programs that encourage men's participation in household work, provide access to affordable childcare, and support flexible work arrangements can help redistribute these responsibilities, reducing women's time poverty and increasing their agency.

Community-level interventions, such as public awareness campaigns and participatory dialogs, are also essential. These initiatives challenge harmful stereotypes, raise awareness about the economic and social benefits of gender equality, and foster community support for women's participation in energy activities. Also, highlighting positive female role models in technical, managerial, and leadership positions can motivate girls and women to pursue energy-related careers. Moreover, involving men and community leaders in awareness campaigns ensures that social change is collective rather than imposed, enhancing the likelihood of sustained impact.

Integrating social and cultural considerations into broader energy programs and policies helps create a supportive ecosystem for women. Mainstreaming gender in energy policy, promoting inclusive community engagement, and coupling technical interventions with gender-transformative initiatives strengthen the systemic changes necessary for women to fully benefit from energy access, technologies, and opportunities. Collectively, these measures address both the visible and invisible barriers rooted in social structures, creating conditions where women's participation in the energy sector can be equitable, meaningful, and sustained.

5.5.7 *Research, Data, and Monitoring*

The research shows a persistent challenge of the limited availability of sex-disaggregated data on employment, career progression, and participation in energy projects. Without robust data, it is difficult to track women's presence, evaluate interventions, or understand the nuanced impacts of gender-inclusive policies (Lazoroska et al., 2024). Research, data collection, and monitoring are essential for evidence-based interventions that promote gender equity in energy systems. It provides the empirical foundation needed for informed decision-making. Without accurate and comprehensive data, it is difficult to identify the barriers women face, measure progress, or design policies and programs that effectively address gender-specific needs.

One critical measure is the collection of sex-disaggregated data on energy access, affordability, and usage. Such data help reveal disparities between men and women in access to electricity, clean cooking fuels, and other energy services. By identifying which populations are underserved or disproportionately burdened by energy poverty, policymakers can design targeted interventions that address these gaps.

Developing gender-sensitive indicators of energy poverty and equity is very important. Standard energy metrics often overlook the nuanced ways gender shapes energy access, household responsibilities, and economic opportunities. Indicators that account for women's time burdens, decision-making power, and participation in energy-related employment or entrepreneurship provide a more comprehensive picture of gendered energy inequalities. These indicators allow policymakers, researchers, and practitioners to monitor progress over time and assess whether interventions are effectively reducing gender disparities.

Finally, assessing the gender impacts of energy projects and policies ensures that energy initiatives do not inadvertently reinforce inequalities. Gender impact assessments can be incorporated at all stages of energy project planning and implementation, from feasibility studies to post-implementation evaluations. These assessments highlight potential barriers women may face and identify opportunities to enhance women's participation.

Addressing gender constraints in energy systems requires a holistic strategy that spans institutional, economic, social, and cultural dimensions. The seven clusters of measures outlined above are interrelated and mutually reinforcing: improving women's access to energy technologies, for example, has limited transformative potential without parallel reforms in policy frameworks, education systems, and social norms. Equally, strengthening data collection and monitoring is essential for ensuring that interventions achieve measurable impacts. Taken together, these measures represent a roadmap for advancing gender equity in energy systems and ensuring that the benefits of the global energy transition are shared more inclusively.

References

Abbas , K .; Xu , D .; Li , S .; Baz , K . Health implications of household multidimensional energy poverty for women: A structural equation modeling technique. Energy and Buildings, 2021, 234, 110661.

Adams, E.A.; Juran, L.; Ajibade, I. 'Spaces of Exclusion' in community water governance: A Feminist Political Ecology of gender and participation in Malawi's Urban Water User. Associations Geoforum, 2018, 95, pp. 133–142.

Alkire, S.; Deneulin, S. "The Human Development and Capability Approach." In An Introduction to the Human Development and Capability Approach: Freedom and Agency, edited by S. Deneulin, and L. Shahani, 2009, pp. 22–48. Routledge, London.

Anderson, S.; Fisher, S. Gender equality and informality in low-carbon transitions. A review of evidence to identify transformative outcomes. IIED, 2022, p. 41. 04-06-2025 Available at: https://www.iied.org/sites/default/files/pdfs/2022-10/21116iied.pdf

Arnaiz, M.; Cochrane, T.A.; Hastie, R.; Bellen, C. Micro-hydropower impact on communities' livelihood analysed with the capability approach. Energy for Sustainable Development, 2018, 45, pp. 206–210.

Baruah B; Gaudet C. Creating and optimizing employment opportunities for women in the clean energy sector in Canada. Journal of Canadian Studies, 2022, 56(2), pp. 240–270.

Buechler, S.; Vázquez-García, V.; Martínez-Molina, K.G.; Sosa-Capistrán, D.M. Patriarchy and (electric) power? A feminist political ecology of solar energy use in Mexico and the United States. Energy Research & Social Science, 2020, 70, 101743.

Campos, F.; Goldstein, M.; McGorman, L.; Munoz Boudet, A.M.; Pimhidzai, O. Breaking the Metal Ceiling: Female Entrepreneurs Who Succeed in Male-Dominated Sectors. Policy Research Working Paper, 2015, No. 7503, p. 39. Available at: https://openknowledge.worldbank.org/server/api/core/bitstreams/844ccb07-4524-5e53-9c91-d1315ba7c0ef/content

CARE. Understanding Gender in Community-Based Adaptation. Practitioner Brief, 2016, 3, p. 24. Available at: file:///C:/Users/gedim/Downloads/CBA-and-Gender-Analysis-Brief3.pdf

Carrere, J.; Peralta, A.; Oliveras, L.; López, M.J.; Marí-Dell'Olmo, M.; Benach, J.; Novoa, A.M. Energy Poverty, Its Intensity and Health in Vulnerable Populations in a Southern European City. Gaceta Sanitaria, 2021, 35, pp. 438–444.

Cellini, M.; Loos, S.; Mirenda, C.; Pisacane, L.; Striebing, C.; Tagliacozzo, S. Exploring the nexus of gender and energy transitions: A systematic literature review. Energy Research & Social Science, 119, 103887.

Crenshaw , K . W . Demarginalizing the intersection of race and sex: A Black feminist critique of antidiscrimination doctrine, feminist theory and antiracist politics. University of Chicago Legal Forum, 1989, 1989(1), 139–167.

EIGE. Gender Equality Action Plans: Method and tool brief. Luxembourg: Publications Office of the European Union, 2024, p. 28. Available at: https://eige.europa.eu/sites/default/files/documents/gender-equality-actions-plans-brief-eige.pdf

EIGE. Gender Equality Training. Gender Mainstreaming Toolkit. Luxembourg: Publications Office of the European Union, 2016, p. 22. Available at: https://eige.europa.eu/sites/default/files/documents/mh0716093enn.pdf

EIGE. Gender Mainstreaming. Gender analysis. Luxembourg: Publications Office of the European Union, 2019, p. 11. Available at: https://eige.europa.eu/sites/default/files/mh0319271enn_002.pdf

EIGE. Gender Budgeting, EIGE, 2017, Vilnius. Available at: https://eige.europa.eu/gender-mainstreaming/methods-tools/gender-budgeting

Energia. Mainstreaming Gender in Energy Projects. A Practical Handbook. 2011, p. 124. Available at: https://energia.org/document/mainstreaming-gender-in-energy-projects-a-practical-handbook/

Hanson, A.; Buechler, S. Towards a Feminist Political Ecology of Women, Global Change and Vulnerable Waterscapes. A Political Ecology of Women, Water and Global Environmental Change, Routledge, London and New York, 2015, pp. 1–16.

Harcourt, W.; Agostino, A.; Elmhirst, R.; Gómez, M.; Kotsila, P. Contours of Feminist Political Ecology. Palgrave Macmillan Cham, 2023, p. 315. Available at: https://doi.org/10.1007/978-3-031-20928-4

Hunt, A.; Samman, E. Women's Economic Empowerment Navigating Enablers and Constraints. Overseas Development Institute, 2016, London, p. 40. Available at: file:///C:/Users/gedim/Downloads/womens-economic-empowerment-odi.pdf

IPCC (Intergovernmental Panel on Climate Change). Climate Change 2023. Synthesis Report. A Report of the Intergovernmental Panel on Climate Change. IPCC, Geneva, Switzerland, p. 184. Available at: https://www.ipcc.ch/report/ar6/syr/downloads/report/IPCC_AR6_SYR_FullVolume.pdf

Johnson, O.W.; Han, J.Y.C.; Knight, A.L.; Mortensen, S.; Aung, M.T.; Boyland, M.; Resurrección, B.P. Intersectionality and energy transitions: A review of gender, social equity and low-carbon energy. Energy Research & Social Science, 2020, 70, 101774.

Kabeer, N. Reversed Realities: Gender Hierarchies in Development Thought. 1994, London; New York: Verso, p. 346.

Kimberle, C. Demarginalizing the Intersection of Race and Sex: A Black Feminist Critique of Antidiscrimination Doctrine, Feminist Theory and Antiracist Politics. 1989, University of Chicago Legal Forum: 1989, 8.

Lazoroska, D.; Palm, J.; Kojonsaari, A.R. Gender-based opportunity structure in the energy sector: a literature review on women's networking and mentoring. Energy, Sustainability and Society, 2024, 14, p. 67.

Moser , C . O . N . Gender planning and development: Theory, practice and training. London: Routledge, 1993, 298 p.

Nguyen, C.P.; Su, T.D. Does Energy Poverty Matter for Gender Inequality? Global Evidence. Energy for Sustainable Development, 2021, 64, pp. 35–45.

Nussbaum, M.C. Creating Capabilities: The Human Development Approach. 2011, Harvard University Press, Belknap Press, p. 256.

Nussbaum, M.C. Women and Human Development: The Capabilities Approach. 2000, Cambridge University Press, p. 312.

OECD. Handbook on the OECD-DAC Gender Equality Policy Marker. 2016, p. 23. Available at: https://canwach.ca/wp-content/uploads/2020/10/Handbook-OECD-DAC-Gender-Equality-Policy-Marker.pdf

Orlando, M.B.; Janik, V.L.; Vaidya, P.; Angelou, N.; Zumbyte, I.; Adams, N. Getting to Gender Equality in Energy Infrastructure: Lessons from Electricity Generation, Transmission, and Distribution Projects. Energy Sector Management Assistance Program (ESMAP) Technical Report, 2018, 012/18. Washington, DC: World Bank. Available at: https://documents1.worldbank.org/curated/en/639571516604624407/pdf/122887-REVISED-GenderEquality-Report-WEB-2-2-18.pdf

Overholt , C .; Anderson , M . B .; Cloud , K .; Austin , J . E . Gender roles in development projects: A case book. West Hartford, CT: Kumarian Press, 1985.

Pearl-Martinez, R., Women at the forefront of the clean energy future, IUCN-USAID, 2014, Washington, DC, p. 55. Available at: https://portals.iucn.org/library/sites/library/files/documents/Rep-2014-005.pdf

Peterson J, Tahssain-Gay L, Benraiss-Noailles L. The impact of exclusivity in talent identification: sources of perceived injustice and employee reactions. Employee Relations, 2022, 44(6), pp. 1217–1240.

Pradhan Shrestha, R.; Mainali, B.; Mokhtara, C.; Lohani, S.P. Bearing the Burden: Understanding the Multifaceted Impact of Energy Poverty on Women. Sustainability 2025, 17, p. 2143.

Pueyo, A.; Maestre, M. Linking energy access, gender and poverty: A review of the literature on productive uses of energy. Energy Research & Social Science, 2019, 53, pp. 170–181.

Rainisio, N.; Boffi, M.; Pola, L.; Inghilleri, P.; Sergi, I.; Liberatori, M. The role of gender and self-efficacy in domestic energy saving behaviors: a case study in Lombardy, Italy. Energy Policy, 2022, 160, p. 112696.

Raman, R.; Ustenko, V.; Filho, W.L.; Nedungadi, P. Energy justice and gender: bridging equity, access, and policy for sustainable development. Discover Sustainability, 2025, 6, p. 558.

Reshi, I.A.; Sudha, T. 2023. "The Gender Pay Gap and Its Impact on Women's Economic Empowerment." Morfai Journal, 3(1), pp. 9–16. Doi.org/10.54443/morfai.v3i1.742

Robeyns, I. Wellbeing, Freedom and Social Justice: The Capability Approach Re-Examined. 2017, Cambridge: Open Book Publishers, p. 266.

Robeyns, I.; Byskov, M.F. "The Capability Approach." In The Stanford Encyclopedia of Philosophy, edited by Edward N. Zalta, and Uri Nodelman. Metaphysics Research Lab, Stanford University, 2020. (Accessed December 28, 2020).

Rojas, A.; Siles, J. Guide on Gender and Energy for Trainers and Managers of Public Policies and Projects. Energia, Olade y Uicn, 2015, p. 160. Available at: https://biblioteca.olade.org/opac-tmpl/Documentos/old0370.pdf

Sadiqa, A.; Sahrakorpi, T.; Keppo, I. Gender vulnerabilities in low carbon energy transitions: a conceptual review. Environmental Research Letters, 2023, 18(4), 043004.

Sen, A. "Equality of what?" In Tanner Lectures on Human Values, 1980, Vol. 1, 197–220. Cambridge: Cambridge University Press.

Shrestha, B.; Tiwari, S.R.; Bajracharya, S.B.; Keitsch, M.M.; Rijal, H.B. Review on the importance of gender perspective in household energy saving behavior and energy transition for sustainability. Energies, 2021, 14(22), p. 7571.

Stephenson, M.A. A Guide to Gender-Responsive Budgeting. Oxfam, Women's Budget Group, 2017, Available at: https://policy-practice.oxfam.org/resources/rough-guide-to-gender-responsive-budgeting-620429/

Streimikiene, D. Low-carbon energy transition from the lens of feminist theories, Contemporary Economics, 2023, 17(4), pp. 456–468.

Sunikka-Blank, M. Gendered roles in household energy consumption: women as absorbers of energy policy. Energy Research & Social Sciences, 2019, 63, 101399.

Therivel , R . Strategic environmental assessment in action, 2nd ed. London: Routledge, 2010, 384 p.

UN Women and UNIDO. Gender Equality and the Sustainable Energy Transition. 2023, p. 64, New York and Vienna. Available at: https://www.unwomen.org/sites/default/files/2023-05/Gender-equality-in-the-sustainable-energy-transition-en.pdf

UN Women. A Gender-Responsive Just Transition For People And Planet. Policy brief, 2023. Available at: https://www.unwomen.org/sites/default/files/2023-11/policy-brief-a-gender-responsive-just-transition-for-people-and-planet-en.pdf

UN Women. Intersectionality resource guide and toolkit. An Intersectional Approach to Leave No One Behind. 2021, p. 55. UN Women Headquarters Office. Available at: https://www.unwomen.org/sites/default/files/2022-01/Intersectionality-resource-guide-and-toolkit-en.pdf

UNDP. Gender Equality Seal. Gender Seal Renewals Guidance Note. 2022, https://www.undp.org/sites/g/files/zskgke326/files/2022-11/undp-ci-gender-seal-renewals-guidance-note-2022.pdf

United Nations Development Programme. Gender Mainstreaming: A Practical Manual for Renewable Energy Initiatives in the Pacific. Gender & Energy Toolkit. 2025, p. 93. Available at: https://www.undp.org/sites/g/files/zskgke326/files/2025-05/undp-gender-mainstreaming-a_practical_manual_for_renewable_energy_initiatives_in_the_pacific-2025.pdf

WFP. Participatory Gender Analysis Report (West Bank & Gaza Strip). 2020, Available at: https://www.un.org/unispal/wp-content/uploads/2020/08/WFPRPT_050820.pdf

Wilgosh B.; Sorman A.H.; Barcena I. When two movements collide: learning from labour and environmental struggles for future just transitions. Futures, 2022, 137, 102903.

World Bank. Women, Business and the Law 2024. 2024, Washington, DC: World Bank. Doi:10.1596/978-1-4648-2063-2.

Wu, Y.; Zhang, L. Evaluation of energy saving effects of tiered electricity pricing and investigation of the energy saving willingness of residents. Energy Policy, 2017, 109, pp. 208–17.

Zhang, Z.; Linghu, Y.; Meng, X.; Yi, H. Is There Gender Inequality in the Impacts of Energy Poverty on Health? Front. Public Health 2022, 10, 986548.

CHAPTER 6

Age-Based Energy Equity

6.1 Age and the Low-carbon Transition

6.1.1 Understanding Age-related Vulnerabilities

The transition from fossil-fuel-based energy systems to low-carbon alternatives has emerged as one of the defining challenges of the twenty-first century, requiring unprecedented coordination across technological, economic, social, and political dimensions while addressing complex questions like equity, justice, and inclusion that extend far beyond technical considerations of energy system design and operation (Creti and Ftiti, 2024). While the urgency of addressing climate change is widely recognized across the world, the social dimensions of energy transitions have received increasing attention from scholars and policymakers just recently. It is necessary to ensure that decarbonization efforts do not exacerbate existing inequalities or create new forms of social exclusion that could undermine both the effectiveness and legitimacy of the low-carbon energy transition (Sovacool et al., 2017).

The recent global crises like COVID-19 pandemic, the Russian-Ukrainian war, Israel-Gaza, and other conflicts pose additional challenges in low-carbon transition justice (Memmott et al., 2021; Żuk, 2023; Agrawal et al., 2024). Among the various dimensions of energy justice that have been identified and analyzed in recent studies, age represents an important but underexamined factor that shapes how different populations experience low-carbon energy transitions, with implications that extend far beyond individual welfare. Age-just, low-carbon energy transition encompasses broader questions of democratic participation, social cohesion, and the effectiveness and sustainability of energy system transformation efforts (Jenkins et al., 2018; Jenkins, 2018). The recognition of age as a critical dimension of energy justice has led to the development of new theoretical frameworks that examine how aging intersects

I. Siksnelyte-Butkiene, D. Streimikiene, *Social Equity in the Energy Sector*,
https://doi.org/10.1007/978-3-032-22224-4_6

with energy systems and energy policy in ways that create both challenges and opportunities for achieving a more inclusive and effective low-carbon energy transition and creating a carbon-neutral society.

The concept of "age-just low-carbon transitions" emerges from the recognition that older adults may face exclusive challenges and barriers in the context of low-carbon energy transitions that require systematic analysis and targeted intervention, while also possessing valuable knowledge, experience, and resources that could contribute to more effective and sustainable energy system transformations. (Żuk and Żuk, 2020). As energy systems become increasingly digitalized, decentralized, and technologically complex, there is growing concern that older adults may be systematically excluded from the benefits of clean energy transitions while bearing high burdens from the costs and disruptions associated with energy system change, creating patterns of age-based inequality that violate principles of energy justice and social equity (Żuk and Żuk, 2022).

This concern has led to the development of new theoretical frameworks, most notably the concept of "energy ageism," which provides a comprehensive lens for understanding how age-based discrimination functions within energy systems and energy policy across multiple interconnected dimensions that create complex patterns of advantage and disadvantage (Żuk and Żuk, 2022). Energy ageism encompasses all socially conditioned mechanisms related to energy policy that disable and discriminate against older people, reducing their quality of life through limited access (economically, technologically, spatially, informationally, and culturally) to different energy sources, new energy technologies, and the benefits of full access to energy. This definition highlights the multidimensional nature of age-based barriers and suggests that achieving a just low-carbon energy transition requires explicit attention to age-related inequalities that operate through complex intersectional mechanisms like economic status, technological access, spatial location, information availability, and cultural recognition.

The significance of addressing age dimensions in low-carbon energy transitions extends beyond concerns about fairness and social justice to encompass practical considerations about the effectiveness and sustainability of decarbonization efforts that depend on broad social participation and support. As noted in a recent paper by Creti and Ftiti, (2024), the low-carbon energy transition will have its winners and losers, and older adults represent a growing proportion of the population in many developed countries, with their specific energy consumption patterns, housing characteristics, economic circumstances, and political participation having important implications for energy system planning, policy design, and democratic legitimacy. Moreover, older adults possess valuable knowledge, experience, and resources that could contribute to the success of the low-carbon energy transition if appropriate mechanisms for inclusion and participation are established. Therefore, the exclusion of older adults represents not only a justice concern but also a missed opportunity for more effective and sustainable energy system transformation.

The demographic context of aging populations in many developed countries makes the integration of age considerations into energy transition planning increasingly urgent and important. Population aging is occurring at an unprecedented pace in many countries, with the proportion of older adults expected to continue growing significantly over the coming decades (World Health Organization, 2015). This demographic transition has important implications for energy systems, as older adults typically have different energy consumption patterns, housing characteristics, and economic circumstances compared to younger age groups, while also facing age-related changes in health, mobility, and social circumstances that can affect their energy needs and their ability to participate in energy system transformations.

The theoretical foundations for understanding age-just low-carbon transitions include energy justice frameworks that have evolved to address distributive, procedural, recognition, and restorative dimensions of energy-related inequality (Heffron, 2022). Energy justice has been increasingly recognized due to its importance of addressing inequities based on income, race, geography, and other factors, but age-based considerations have received less systematic attention despite growing evidence of their significance for understanding how energy systems affect different populations and how energy transitions can be designed to promote rather than undermine social equity and inclusion (Jenkins et al., 2016; Jenkins, 2018).

The integration of aging problems with energy research has revealed complex patterns of disadvantage that require sophisticated analytical frameworks capable of addressing multiple intersecting forms of inequality simultaneously (Calasanti, 2015). Contemporary aging studies have documented extensive evidence of age-based discrimination across multiple domains of social life, including employment, healthcare, housing, and social services, while also highlighting the diversity of older adult experiences and the importance of avoiding ageist stereotypes that assume all older adults have similar needs, preferences, or capabilities (Ayalon and Tesch-Römer, 2018).

Contemporary energy justice frameworks typically organize analysis around four key tenets of energy inequalities: distributive justice, which focuses on how energy benefits and burdens are distributed across different populations; procedural justice, which examines who participates in energy decision-making processes and how those processes are structured; recognition justice, which addresses whose needs, values, and ways of life are acknowledged and respected in energy systems; and restorative justice, which focuses on addressing past harms and preventing future ones (Sovacool et al., 2017; Sovacool, 2021). Each of these tenets offers important insights for understanding how age intersects with energy systems and how energy ageism might be addressed through comprehensive policy interventions that address multiple issues of disadvantage simultaneously.

The empirical evidence for age-related energy inequities has grown substantially in recent years, with comparative studies revealing significant variations in how different welfare state models and policy approaches affect older adults' low-carbon energy transition experience (Szulecki et al., 2024). The spatial and social determinants of energy ageism have also received increasing attention in scientific works and showed that the type of building and its location within urban space serve as significant determinants of energy poverty among older adults, creating complex patterns of spatial inequality that intersect with age-related vulnerabilities in ways that require comprehensive policy responses (Żuk and Żuk, 2024a). Therefore, it is important to understand how built environment factors, housing characteristics, and geographic location interact with age-related vulnerabilities that require comprehensive policy responses addressing multiple scales of intervention from individual households to communities, cities, and regions.

The policy implications of this emerging research are far-reaching, suggesting the need for fundamental changes in how energy systems are designed, regulated, and operated to ensure age inclusion across multiple dimensions of energy access, participation, and benefit distribution (Heffron, 2022). This includes not only targeted programs and services for older adults but also broader changes in energy governance frameworks, regulatory approaches, and institutional practices for addressing energy ageism while also leveraging older adults' capabilities and contributions to low-carbon energy transition. The development of age-inclusive energy policies requires understanding the complex interactions between individual-level factors such as health status and economic resources, household-level factors such as housing characteristics and family composition, community-level factors such as social capital and local infrastructure, and broader structural factors such as welfare state institutions, energy market design, and cultural attitudes toward aging.

6.1.2 Conceptualizing "Energy Ageism" in the Transition

6.1.2.1 The Emergence and Conceptualization of Energy Ageism

The concept of energy ageism represents a significant theoretical innovation in the intersection of aging studies and the energy research field, building on established scholarship on ageism in other domains while providing a framework specifically tailored to understanding how age-based discrimination operates within low-carbon energy transitions and energy policy in ways that create systematic patterns of disadvantage and exclusion for older adults (Żuk and Żuk, 2022). Table 6.1 presents several definitions of energy ageism by providing the main focus of the concept based on recent scientific studies.

The table above provides a comprehensive overview of the concept of "energy ageism" by compiling definitions from the main academic papers linked to ageism and contextualizing them. The information provided in the table shows how energy ageism is defined both as a structural issue of systemic discrimination within energy policy and as a lived experience of hardship related to energy poverty and exclusion from the low-carbon energy transition of older adults. By including general definitions of ageism, the table illustrates how the specific challenges faced by older people in the energy sector are

Table 6.1 Definitions of energy ageism. Source: created by the author based on Butler (1969), World Health Organization (2015), Żuk, and Żuk (2022, 2024a,b), and Levin and Nevo (2024).

Definition of energy ageism	*Focus*	*Source*
All socially conditioned mechanisms related to energy policy that disable and discriminate against older people, reducing their quality of life through limited (economically, technologically, spatially, informationally, and culturally) access to different energy sources, new energy technologies, and the benefits of full access to energy supply.	This broad definition follows a holistic discrimination approach and sees energy ageism as a structural problem where social and policy mechanisms systematically block seniors from participating in and benefiting from the modern energy system.	Żuk and Żuk, 2022
Energy ageism covers all the mechanisms of discrimination against seniors related to energy poverty and problems with paying energy and heating bills, a lack of support for older people during the low-carbon energy transition, forcing them to stay in cold and underheated homes, as well as the cultural and technological barriers to accessing new energy sources in their everyday life.	This definition focuses on the practical, everyday consequences of energy ageism, linking it directly to energy poverty, lack of support, and the physical reality of living in cold homes.	Żuk and Żuk, 2024a,b
An age-based pattern of exclusion conceptualized in the context of decarbonization.	This frames energy ageism as a specific form of injustice that can occur during the societal shift away from fossil fuels, where age becomes a key dimension of exclusion from energy decarbonization.	Szulecki et al., 2024;Żuk and Żuk, 2024a
Ageism refers to the stereotypes (how we think), prejudice (how we feel), and discrimination (how we act) towards others or oneself based on age.	This general ageism definition from the WHO is not specific to energy but provides the three core dimensions—stereotypes, prejudice, and discrimination—that underpin the concept of energy ageism. It applies to all age groups.	World Health Organization, 2015
A combination of three connected elements: prejudicial attitudes towards older people and the aging process; discriminatory practices against older people; and institutional practices and policies that perpetuate stereotypes about elderly people.	This classic definition breaks down ageism into three interconnected pillars of ageism: negative attitudes, discriminatory actions, and institutional policies that reinforce stereotypes.	Butler, 1969
A related concept, eco-ageism, is defined as the reduction of older people's relations with nature to the stereotypes of weakness, irrelevance, and passivity.	While distinct, this concept is relevant as energy ageism, as it focuses on the socioeconomic and political system's discriminatory mechanisms during the low-carbon energy transition.	Levin and Nevo, 2024

rooted in larger societal patterns of stereotyping, prejudice, and discriminatory practices based on age.

Table 6.2 presents the description of the concept of energy ageism with key manifestations and policy implications. Also, it provides a structured overview of the concept, breaking it down into four key thematic areas for clarity. It begins with a core definition of energy ageism as systemic discrimination and then systematically explores its causes and consequences by detailing its socio-economic, spatial/architectural, and institutional determinants. For each category, the table offers concrete examples—such as low pensions, poor housing insulation, and bureaucratic barriers—to illustrate how these factors manifest in the real world, while also outlining corresponding policy solutions designed to mitigate these specific challenges and promote a more just and inclusive low-carbon energy transition.

The concept of energy ageism has emerged from recognition that existing approaches to energy justice, while increasingly sophisticated in their treatment of various forms of inequality, had not adequately addressed age as a dimension of energy-related disadvantage that requires systematic analysis and targeted intervention beyond general approaches to social equity and inclusion (Jenkins, 2018).

The theoretical development of energy ageism draws on broader scientific studies linked to institutional discrimination and systemic inequality, suggesting that age-based disadvantage in energy systems is not simply the result of individual prejudices or biological aging processes but reflects deeper patterns of social organization and resource allocation that are embedded within the structures, technologies, and practices of energy systems themselves (Fuller and McCauley, 2016). This structural understanding of energy ageism represents a departure from individualistic approaches to aging that focus primarily on personal characteristics or capabilities, instead emphasizing the social, economic, and political factors that shape how aging intersects with energy access and participation in ways that can be modified through appropriate policy interventions and institutional changes.

The conceptual foundations of energy ageism are rooted in broader theoretical work on ageism that has documented how age-based stereotypes, prejudices, and discrimination operate across multiple domains of social life, including employment, healthcare, housing, and social services (Butler, 1969). Early studies on ageism focused primarily on individual attitudes and behaviors, but more recent work has emphasized the institutional and structural dimensions of age-based discrimination that operate through policies, practices, and organizational arrangements that systematically disadvantage older adults regardless of individual intentions or attitudes (Levy and Banaji, 2002).

Contemporary ageism studies have also emphasized the intersectional nature of age-based discrimination, recognizing that ageism intersects with other forms of inequality based on gender, race, class, sexuality, and other factors in ways that create complex patterns of disadvantage that cannot be understood by examining age in isolation (Krekula et al., 2018). This intersectional

Table 6.2 Key manifestations of energy ageism and policy implications. Source: created by the author based on Żuk and Żuk, (2020; 2022, 2024a, b).

Category	*Description*	*Key manifestations*	*Policy implications*
Definition	Systemic discrimination against older people in the context of energy policy and low-carbon energy transition, limiting their access to affordable, clean energy and reducing their quality of life.	Seniors are being unable to afford heating bills due to their low pension. Exclusion from subsidy programs due to digital illiteracy or lack of upfront funds for seniors. Seniors are being forced to live in cold, unhealthy conditions.	Integrating social justice into all energy transition policies; Creating specific legal and social protections for seniors.
Socioeconomic Determinants	Factors related to income, class, and financial resources that create vulnerability.	Fixed pensions are insufficient to cover rising energy prices and renovation costs.Seniors must choose between heating, food, and medicine.	Reforming the pension system to guarantee a decent standard of living.Developing the targeted financial support programs that cover 100% of costs for low-income seniors.
Spatial and Architectural Determinants	Factors related to housing type, location, and physical infrastructure.	Poor insulation, high ceilings, and lack of modern heating systems in old buildings where older adults reside.No access to district heating or gas networks in certain urban and rural areas populated by older adults. Traditional rural homes populated by older adults often rely on coal, while suburban homes have modern systems.	Investing in public infrastructure to connect older districts to efficient heating networks.Funding systemic thermal modernization of entire neglected housing quarters. Adapting solutions to the specific needs of urban vs. rural areas.
Institutional and Policy Determinants	Barriers created by government policies, bureaucracy, and lack of support from local authorities.	Complex, online-only application forms for furnace replacement grants create bureaucratic hurdles for seniors.Anti-smog regulations are imposed without providing viable financial alternatives. Local housing authorities often block grassroots initiatives for renewable energy (e.g., solar panels).	Simplifying bureaucracy and providing in-person assistance for grant applications.Empowering local authorities to support and fund resident-led energy initiatives. Mandating public institutions (schools, clinics) to lead the way in adopting renewable energy.

(continued)

Table 6.2 (continued)

Category	*Description*	*Key manifestations*	*Policy implications*
Consequences and Impacts	The negative outcomes of energy ageism on individuals and society.	Increased respiratory illnesses and poor mental health due to cold homes and financial stress of older adults Worsened air quality from burning low-quality fuel or waste. Deepened distrust in government and new technologies; increased social inequality by older adults	Linking public health policies with energy poverty reduction strategies. Creation of educational campaigns to build trust and ecological awareness. Fostering of intergenerational solidarity and community-based energy projects.

understanding of ageism provides important insights for understanding energy ageism, as older adults' experiences of energy systems are shaped not only by their age but also by their economic circumstances, housing situations, health conditions, geographic locations, culture, and other factors that interact with age in complex ways.

Żuk and Żuk's (2022) foundational work on energy ageism identifies it as a form of systemic discrimination that operates through multiple interconnected mechanisms that create barriers to older adults' full participation in low-carbon energy transitions across five key dimensions: economic, technological, spatial, informational, and cultural. The theoretical framework is grounded in understanding of the relationship between biological aging processes and social structures, recognizing that while aging involves certain biological changes that may affect individuals' energy needs and capabilities, the significance and impact of these changes are largely determined by social, economic, and political factors that can be modified through appropriate policy interventions and institutional reforms.

The concept of energy ageism operates across five key dimensions that create barriers to older adults' full participation in energy systems, each of which requires detailed analysis to understand how they interact to create complex patterns of disadvantage that compound over time and across different domains of experience. Economic barriers encompass the financial constraints that may limit older adults' ability to invest in new energy technologies, upgrade their homes for energy efficiency, or access premium energy services. These economic barriers are particularly significant given that many older adults live on fixed incomes and may face higher energy costs due to spending more time at home, having health conditions that require additional energy use for heating, cooling, or medical equipment, or living in older, less energy-efficient housing that requires more energy to maintain comfortable and healthy living conditions (Żuk and Żuk, 2024a).

Technological barriers reflect the increasing digitization, use of Artifical Inteligence (AI) and complexity of modern energy systems, which may create challenges for older adults who have had less exposure to digital technologies or who may face age-related changes in vision, hearing, or cognitive processing that affect their ability to navigate complex technological interfaces, understand new energy technologies, or adapt to changing energy system requirements (Axon and Morrissey, 2020; Bouzarovski et al., 2023). However, it is important to note that the theoretical framework of energy ageism explicitly rejects ageist assumptions about older adults' low technological capabilities, instead focusing on how energy technologies are designed and implemented in ways that may not accommodate diverse user needs and preferences, suggesting that technological barriers are largely the result of bad design or implementation practices rather than inherent limitations of older adults.

Spatial barriers encompass the geographic and built environment factors that may affect older adults' access to modern energy services and technologies, including the concentration of older adults in certain types of housing such as older, less energy-efficient buildings, geographic areas with limited access to energy infrastructure or services, and mobility limitations that may affect older adults' ability to access energy-related services or participate in community-energy programs (Żuk and Żuk, 2024b). The spatial dimension of energy ageism is particularly important because it highlights how historical patterns of urban development and housing construction interact with contemporary energy system transitions to create uneven geographies of opportunity and disadvantage that may systematically disadvantage older adults.

Informational barriers refer to challenges in accessing, understanding, and acting on energy-related information, which become increasingly significant as contemporary energy systems become more complex and energy markets more competitive, requiring consumers to navigate increasingly sophisticated choices about energy suppliers, technologies, and services (Szulecki et al., 2024). Information about energy programs and services may be disseminated primarily through digital channels that some older adults may not regularly access or may be presented in formats that do not accommodate age-related changes in information processing, vision, or hearing. The informational dimension of energy ageism also encompasses broader issues of energy literacy and the capacity to navigate increasingly complex energy choices that require specialized knowledge and skills.

Cultural barriers represent perhaps the most subtle but pervasive dimension of energy ageism, encompassing the attitudes, assumptions, and practices within energy institutions and communities that marginalize or exclude older adults from energy decision-making processes, program design, and service delivery (Fuller and McCauley, 2016). These cultural barriers may manifest in energy policy discussions that implicitly assume younger, more technologically savvy consumers, leading to program designs that do not consider older adults' needs or preferences. For example, community-energy initiatives usually inadvertently exclude older adults through meeting formats, communication styles, or decision-making processes that do not accommodate diverse participation styles and are not in line with older adult's preferences.

6.1.2.2 *The Vicious Circle of Energy Ageism and Intersectional Disadvantage*

One of the most significant theoretical insights from the energy ageism framework is its identification of self-reinforcing cycles of exclusion and disadvantage that operate across multiple dimensions simultaneously, creating cumulative patterns of disadvantage that become increasingly difficult to address over time. Żuk and Żuk (2022) describe this as a "vicious circle" in which the previously existing hierarchies inscribed in the economic, cultural, and spatial dimensions of the system are strengthened due to the weak solidarity of the social and energy policy between the generations. This conceptualization highlights how energy ageism operates not as a simple additive effect of individual barriers but as a complex system of interconnected disadvantages that compound over time and across different domains of experience, creating increasingly embedded patterns of exclusion that require comprehensive and sustained policy interventions and institutions to address effectively.

The vicious circle of energy ageism operates through multiple pathways that compound age-based disadvantages over time and create increasingly difficult barriers to overcome, with implications that extend beyond individual older adults to entire communities and regions where older adults are concentrated (Table 6.3). Older adults who are excluded from initial low-carbon energy transition programs may fall further behind as energy systems continue to evolve, making it increasingly difficult for them to catch up with technological and policy changes that assume familiarity with previous innovations and that build on earlier investments and experiences. This creates a form of path dependency in which early exclusion from the low-carbon energy transition leads to cumulative disadvantage that becomes increasingly difficult to address through later interventions and policy mechanisms.

Table 6.3 breaks down the self-perpetuating cycle of energy ageism into four distinct stages, explaining the mechanism at each step and providing the outcome that leads to the next stage of the vicious cycle.

Table 6.3 systematically breaks down the vicious cycle of energy ageism into four distinct, sequential stages to clarify how systemic discrimination becomes self-perpetuating. It begins by outlining the foundational hostile societal and institutional conditions that disadvantage older people, which in turn leads to their weakened position and reduced role within the low-carbon energy transition. Then it details how this marginalization results in unfavorable transition effects, where seniors bear the costs but miss the benefits. Finally, Table 6.3 explains the crucial reinforcement stage, where these negative outcomes are used to justify the initial discriminatory framework, thus completing the loop and ensuring the cycle repeats with even greater intensity again and again.

Economic disadvantages may compound over time as older adults miss opportunities for energy savings or efficiency improvements that could reduce

Table 6.3 The vicious circle of energy ageism. Source: created by the author based on Żuk and Żuk (2022, 2024a,b), Furszyfer Del Rio et al. (2021), and Ciplet (2021).

Stage	*Description*	*Outcome*
1. Hostile Societal and Institutional Conditions	The cycle begins with a societal framework that is indifferent or hostile to the needs of older people. This includes inadequate legal, economic, and social systems that fail to provide a sufficient safety net.	Seniors are systematically disadvantaged due to the lack of support and discriminatory policies that directly weaken their social and economic position.
2. Weakened Position and Reduced Role of Seniors	As a direct result of the hostile conditions, older people lack the financial resources and agency to participate in the low-carbon energy transition. They cannot afford to invest in modern heating or insulation and are excluded from decision-making.	Seniors are excluded from the benefits due to the inability of the older population to participate, and as they are disproportionately burdened by the costs and negative aspects of the low-carbon energy transition.
3. Unfavorable Transition Effects for Seniors	Because older adults are unable to adapt or participate, the outcomes of the low-carbon energy transition are skewed against them. They are left with higher energy costs, poor living conditions, and no access to the benefits of cleaner, more efficient energy.	Negative stereotypes appear to be confirmed; that is, the poor outcomes are misinterpreted not as a result of systemic failure but as a result of the seniors' own "inability" or "unwillingness" to change.
4. Reinforcement of Discrimination and Hostility	The unfavorable outcomes are used, implicitly or explicitly, to justify the initial hostile framework. Policymakers see no reason to change a system for a group that appears "passive" or "resistant," thus reinforcing the stereotypes and relevant policies.	The cycle is completed and strengthened as the institutional hostility is now more entrenched, ensuring that the next wave of policies will also be discriminatory, starting the cycle all over again.

their long-term energy costs, creating a situation where those who most need financial relief are least able to access the programs and technologies that could provide it (Żuk and Żuk, 2024a, b). This economic dimension of the vicious circle is particularly problematic because it creates a form of regressive distribution in which energy transition benefits flow disproportionately to those who already have greater economic resources, while those with the greatest need for energy cost reductions are systematically excluded from accessing these benefits.

Technological barriers may become more pronounced as energy systems become increasingly digitized and automated, creating growing gaps between those who can navigate digital-energy systems and those who cannot, with implications that extend beyond simple access to energy services to access to information about energy programs and policies and the ability to optimize energy consumption through smart technologies and demand response programs (Furszyfer Del Rio et al., 2021). This digital divide in energy systems has implications that extend beyond individual disadvantage to affect the overall effectiveness of low-carbon energy system transformation. The exclusion of the older adults, making a significant share of the population, from digital-energy systems may limit the potential for demand-side management and other efficiency improvements that depend on broad participation in these programs.

The intergenerational dimension of this vicious circle is particularly important for understanding how energy ageism operates within broader social and political contexts that shape energy policy and energy system development (Jenkins et al., 2018). Low-carbon energy transitions are often framed in terms of intergenerational justice, scarifying benefits for future generations, but this framing can inadvertently marginalize older adults by suggesting that their current needs and preferences are less important than those of younger generations. This can lead to energy policies that prioritize long-term environmental benefits over short-term accessibility and affordability concerns that disproportionately affect older adults.

The intersectional nature of energy ageism is also crucial for understanding how age-related disadvantages interact with other forms of inequality based on income, race, gender, geography, health status, and other factors that create complex patterns of multiple disadvantage (Ciplet, 2021). Older adults who experience multiple forms of disadvantage may face particularly severe barriers to energy access and participation, while those with greater resources may be better positioned to navigate energy system changes and access all available benefits of the low-carbon energy transition. This intersectional understanding suggests that effective responses to energy ageism must address multiple dimensions of disadvantage simultaneously rather than treating age as an isolated factor. It requires a comprehensive approach that addresses the full range of factors that affect energy equity (Goddard and Farrelly, 2018).

The spatial dimensions of disadvantage are particularly important for understanding how energy ageism operates in practice, as older adults are not evenly distributed across geographic areas but tend to be concentrated in certain neighborhoods, housing types, and regions that may have obsolete energy infrastructure, poor services, and fewer opportunities (Żuk and Żuk, 2024a, b). These spatial concentrations of older adults may create additional challenges for addressing energy ageism, as they may require coordinated interventions that address neighborhood-level or regional factors rather than simply individual household characteristics.

The analysis of age-just, low-carbon energy transitions reveals the emergence of a significant new area of scholarship and policy concern that sits at the intersection of aging studies, energy research, and social justice, with important implications for understanding how energy system transformations can be designed and implemented to promote rather than undermine age equity while also achieving ambitious climate goals. The concept of energy ageism provides a valuable framework for understanding how age-based discrimination operates within energy systems and energy policy, while the broader framework of age-just transitions offers a lens for examining how energy system transformations can be designed and implemented to promote rather than undermine age equity across multiple dimensions of access, participation, and benefit distribution.

The theoretical contributions of this emerging field are substantial, providing new insights into the mechanisms through which age-based disadvantage operates in energy contexts and the ways that energy transitions may create or exacerbate age-related inequalities if not carefully designed and implemented with explicit attention to age inclusion. The concept of energy ageism, with its identification of economic, technological, spatial, informational, and cultural barriers, provides a comprehensive framework for analyzing age-related energy disadvantage that goes beyond simple assumptions about older adults' capabilities or preferences to address the structural and institutional factors that create and maintain age-based energy inequities.

6.2 Theoretical and Methodological Perspectives

6.2.1 Energy Justice Frameworks and Age Dimensions

The theoretical development of energy ageism has occurred within the broader context of energy justice scientific literature, which provides important conceptual tools and normative frameworks for understanding and addressing energy-related inequities across multiple dimensions and scales of analysis (Sovacool et al., 2017). Energy justice, as defined by leading scholars in this field, encompasses "a global energy system that fairly distributes both the benefits and burdens of energy services, and one that contributes to more representative and inclusive energy decision-making" (Heffron, 2022). This definition highlights both distributive and procedural dimensions of energy justice while also emphasizing the importance of inclusive participation in energy governance and decision-making processes.

Table 6.4 presents energy justice frameworks. These frameworks provide different lenses through which it is possible to understand and address inequalities in the energy system.

Table 6.4 provides a concise summary of the primary theoretical frameworks used to analyze and address energy justice. It systematically outlines six distinct but interrelated concepts of energy justice: distributive, procedural, recognition, restorative, cosmopolitan, and energy sovereignty. For each framework, the table highlights its core tenets, poses the key questions it seeks to answer, and provides an academic reference that established the concept.

Therefore, the contemporary scientific literature on energy justice is typically organized around four key tenets that provide a comprehensive framework for analyzing and addressing energy inequities: distributive justice, procedural justice, recognition justice, and restorative justice (Sovacool et al., 2017). Each of these tenets offers important insights for understanding how age intersects with energy systems and how energy ageism might be addressed through comprehensive policy interventions that address multiple dimensions of disadvantage simultaneously of older adults.

Table 6.4 Summary of energy justice frameworks and their focus. Source: created by the author based on Shrader-Frechette (2002), Schlosberg (2007). Endres (2009), Healy and Barry (2017), Walker (2012), Sovacool and Dworkin (2015), Sovacool et al. (2017), and Jenkins et al. (2016).

Framework	*Focus*	*Key questions*	*References*
Distributive Justice	This framework focuses on the equitable distribution of benefits and burdens of the low-carbon energy transition. It examines who gets the "goods" (e.g., clean energy, green jobs) and the "bads" (e.g., air pollution, high energy costs).	Are the costs and benefits of energy systems shared fairly across all social groups? Are certain communities disproportionately burdened by energy infrastructure or high energy prices?	Shrader-Frechette, 2002; Schlosberg, 2007
Procedural Justice	This framework addresses the fairness in the decision-making process. It emphasizes inclusive, transparent, and accountable governance where all affected communities have a meaningful voice in energy-related decisions.	Do all stakeholders have an opportunity to participate in energy-related decisions?Are decision-making processes transparent and are officials accountable? Is information accessible to all?	Walker, 2012; Healy and Barry, 2017
Recognition Justice	This framework focuses on acknowledging and respecting the rights, cultures, and unique vulnerabilities of different social groups, particularly those who have been historically marginalized. It is about recognizing diverse identities and experiences.	Are the unique values, cultures, and needs of marginalized communities recognized and respected in energy policy?Are the harms experienced by specific groups (e.g., Indigenous communities, the elderly) acknowledged?	Schlosberg, 2007; Sovacool and Dworkin, 2015
Restorative Justice	This framework aims to repair the harm done to communities that have been historically and disproportionately burdened by past energy systems (e.g., from fossil-fuel extraction and air pollution). It seeks to correct past injustices.	What measures can be taken to repair the damage done to communities harmed by past energy practices? How can we remediate polluted sites and compensate affected populations?	Jenkins et al., 2016
Cosmopolitan Justice	This framework extends the principles of energy justice beyond national borders to a global scale. It considers the responsibilities that nations have to one another, especially regarding climate change and resource distribution.	How can energy justice be achieved globally, across different nations and generations?What are the duties of high-emitting countries to those most vulnerable to climate change?	Sovacool et al., 2017

Table 6.4 (continued)

Framework	*Focus*	*Key questions*	*References*
Energy Sovereignty	A framework, often advanced by Indigenous communities that asserts the right of communities to control their own energy systems—from generation and distribution to consumption—in a way that aligns with their cultural, economic, and ecological values.	Do communities have the right to own and control their own energy resources and infrastructure? How can energy systems be designed to support local self-determination and cultural values?	Endres, 2009

Distributive justice focuses on how energy benefits and burdens are distributed across different populations, with particular attention to ensuring that all people have access to affordable, reliable energy services while minimizing harmful impacts on any particular group (Heffron, 2022). When applied to age considerations, distributive justice highlights several key concerns about how energy costs, benefits, and impacts are distributed across age groups, with research consistently showing that older adults face higher energy burdens than younger adults, particularly those with lower incomes who may spend a disproportionate share of their income on energy costs while also facing barriers to accessing energy efficiency improvements and clean energy technologies that could reduce their long-term energy costs (Eaton, 2021).

Research on energy burden consistently shows that older adults face higher energy costs than younger adults, reflecting both lower incomes and higher per-capita energy consumption among older adults who may spend more time at home and have health conditions that require additional energy use for heating, cooling, or medical equipment. However, distributive justice analysis must also consider the diversity of older adult experiences, as energy burdens vary significantly based on housing type, geographic location, health status, and other factors that create complex patterns of disadvantage within older adult populations that require nuanced policy responses.

Distributive justice concerns also extend to access to energy system benefits, such as renewable energy installations, energy efficiency programs, and distributed-energy resources that can provide both environmental and economic benefits while also contributing to energy system resilience and sustainability (Cha, 2020; Goddard and Farrelly, 2018). Older adults may face barriers to accessing these benefits due to economic constraints, technological challenges, or program design features that do not accommodate their needs and preferences. For example, solar panel installation programs may require significant upfront investments that are problematic for older adults having low fixed incomes or may involve complex financing arrangements that are difficult to navigate without appropriate support and assistance.

The temporal dimensions of distributive justice are also important for understanding age-related energy concerns, as energy investments and policies have different time horizons that may affect different age groups differently (Jenkins et al., 2018). Energy efficiency improvements and renewable energy installations typically require significant upfront investments that provide benefits over many years, which may create challenges for older adults who may not expect to remain in their current homes long enough to recoup these investments. This temporal mismatch between investment requirements and benefit realization may create systematic barriers to older adult participation in low-carbon energy transition programs that require innovative financing mechanisms and policy designs to address effectively.

Procedural justice examines who participates in energy decision-making processes and how those processes are structured, with particular attention to ensuring that all affected parties have meaningful opportunities to participate in decisions that affect them (Sovacool et al., 2017). This includes not only formal policymaking processes but also program design, implementation, and evaluation activities that shape how energy systems operate and evolve over time. Age-related procedural justice concerns include the potential exclusion of older adults from energy planning and decision-making processes due to barriers such as meeting formats that are not accessible, communication strategies that do not reach older adult communities, or decision-making processes that do not accommodate diverse participation styles and preferences.

Traditional public participation methods such as evening meetings, online surveys, or technical workshops may not be accessible to all older adults due to mobility limitations, digital barriers, or communication preferences that differ from those assumed by energy planners and policymakers (Szulecki et al., 2024). Additionally, energy decision-making processes may implicitly prioritize perspectives and concerns that are more common among younger adults, such as long-term environmental benefits or technological innovation, while giving less attention to concerns that may be more important for older adults, such as affordability, reliability, and ease of use.

However, procedural justice also recognizes the valuable perspectives and experiences that older adults can bring to energy decision-making, including longer-term perspectives on energy system development, experience with different energy technologies and practices, and insights into community needs that can inform more effective energy policies and programs (Fuller and McCauley, 2016). Ensuring procedural justice requires creating multiple pathways for older adult participation and recognizing the diverse ways they can contribute to energy decision-making processes, including not only formal participation in energy planning processes but also informal knowledge sharing, education, and community organizing activities.

Recognition justice addresses whose needs, values, and ways of life are acknowledged and respected in energy systems, recognizing that different groups may have different relationships with energy, different priorities and preferences, and different ways of understanding and engaging with energy issues (Sovacool et al., 2017). Recognition justice requires moving beyond one-size-fits-all approaches to energy policy and programming to acknowledge and accommodate this diversity in ways that respect the dignity and autonomy of all energy system participants while also recognizing the valuable contributions that different groups can make to energy system transformation.

From an age perspective, recognition justice highlights the importance of acknowledging the diversity of older adult experiences and avoiding ageist stereotypes that assume all older adults have similar needs, preferences, or capabilities (Cebola et al., 2023). It also requires recognizing the valuable knowledge and perspectives that older adults bring to energy systems, including traditional knowledge about energy conservation, experience with different energy technologies, and insights into long-term energy system changes necessary for effective sustainable energy policies (Willand, 2022). Recognition justice also addresses the tendency to frame low-carbon energy transition primarily in terms of benefits for future generations, which can inadvertently marginalize older adults by suggesting that their current needs and preferences are less important (Stephens, 2019).

A recognition justice approach would acknowledge that energy systems must serve current older adults effectively while also preparing for the needs of future generations and that older adults have legitimate interests in energy system design and operation that must be respected and accommodated in energy planning and policymaking processes (Jenkins, 2018; Jenkins et al., 2018). This includes recognizing older adults' preferences for energy system characteristics such as reliability, simplicity, and affordability, while also acknowledging their potential contributions to energy conservation, community resilience, and social sustainability that may be overlooked in technology-focused approaches to low-carbon energy transitions.

Restorative justice focuses on addressing past harms and preventing future ones, including addressing the legacy impacts of harmful energy practices, compensating communities that have borne disproportionate energy burdens, and ensuring that energy transitions do not perpetuate or create new forms of harm (Heffron, 2022). Age-related restorative justice concerns include addressing the ways that past energy policies and practices may have disadvantaged older adults, such as through exposure to air pollution from fossil-fuel facilities and exclusion from energy efficiency programs that could have provided long-term benefits.

Restorative justice also includes ensuring that current energy transitions do not create new forms of age-based harm or exclusion, which might involve targeted programs to help older adults access energy efficiency improvements, compensation for communities that have borne disproportionate air pollution burdens over time, or proactive measures to prevent low-carbon energy transitions from exacerbating existing age-related disadvantages (Żuk and Żuk, 2024a, b). The restorative dimension of energy justice is particularly important for addressing the aggregate effects of energy-related disadvantages that may have accumulated over older adults' lifetimes and that may require comprehensive interventions to address them effectively.

6.2.2 Integrating Energy Ageism into Equity Frameworks

The integration of energy ageism concepts with broader energy justice frameworks provides a more comprehensive understanding of how age intersects with energy systems and energy policy while offering tools for developing more effective interventions that address multiple dimensions of disadvantage simultaneously (Tikkakoski et al., 2024). Energy ageism can be understood as a violation of energy justice principles, representing failures of distributive, procedural, recognition, and restorative justice as they apply to age-related concerns, while energy justice frameworks provide systematic approaches for analyzing age-related energy concerns and identifying potential interventions that can address multiple dimensions of disadvantage simultaneously.

The four tenets of energy justice offer a systematic way of analyzing age-related energy concerns and identifying potential interventions that address multiple dimensions of disadvantage simultaneously while also leveraging older adults' capabilities and contributions to energy system transformation. The emphasis on both outcomes and processes highlights the importance of not only ensuring that older adults have access to modern energy services but also that they can participate meaningfully in energy decision-making processes that affect their and communities lives (Heffron, 2022). This dual focus on outcomes and processes is particularly important for addressing energy ageism because it recognizes that exclusion from decision-making processes can be as harmful as exclusion from energy services themselves.

The integration of these frameworks highlights the interconnected nature of different forms of inequality and the need for comprehensive approaches that address multiple dimensions of disadvantage simultaneously (Ciplet, 2021). Energy ageism intersects with other forms of inequality based on income, race, geography, health status, and other factors, requiring intersectional approaches that address multiple forms of energy inequality simultaneously rather than treating age as an isolated factor. This intersectional understanding is crucial for developing effective interventions because it recognizes that older adults' energy experiences are shaped by multiple factors that interact in complex ways (Fenger, 2007).

Furthermore, the integration of energy ageism and energy justice frameworks provides a foundation for developing more inclusive and effective energy policies and programs that can advance both environmental and social goals simultaneously (Jenkins et al., 2016). By explicitly considering age dimensions within energy justice frameworks, policymakers and practitioners can develop interventions that address the full range of barriers that older adults face while also leveraging their capabilities and contributions to energy system transformation. This integration also provides a framework for evaluating the effectiveness of energy policies and programs in addressing age-related concerns and for identifying areas where additional interventions may be needed.

In Table 6.5 the integration of energy justice with energy ageism is provided.

Table 6.5 The integration of energy ageism in energy justice frameworks. Source: created by the authors based on Walker (2012), Żuk and Żuk (2024a), Jenkins et al. (2016), Schlosberg (2007), and Żuk and Żuk (2022).

Energy justice frameworks	*Application to energy ageism*	*Key questions for an age-inclusive low-carbon energy transition*	*References*
Distributive Justice	The unfair distribution of energy costs and benefits. Energy ageism manifests as seniors bearing the burdens of high energy costs, cold homes, and poor air quality, while being excluded from the benefits of modern, efficient energy systems.	Are seniors disproportionately paying more of their income for energy? Do the benefits of energy efficiency programs and subsidies reach older, low-income households?	Walker, 2012; Żuk and Żuk, 2024a
Procedural Justice	Unfairness in decision-making processes. Energy ageism is evident when policies are created without consulting older adults. Bureaucratic hurdles, digital-only applications, and complex language effectively bar older adults from participating in and accessing support programs.	Are older adults and their advocates included in the planning of energy policies?Are support programs designed to be accessible for those without digital literacy or upfront capital?	Jenkins et al., 2016
Recognition Justice	The failure to acknowledge and respect a group's unique status and needs. Energy ageism is a failure to recognize seniors as a distinct group with specific vulnerabilities (e.g., fixed incomes, health issues, lower mobility) and unique strengths (e.g., life experience). Their needs are often rendered invisible by policymakers.	Do energy policies acknowledge the unique financial and health vulnerabilities of older people?Are seniors seen as passive recipients of policy or as active agents whose needs and rights must be respected?	Schlosberg, 2007; Żuk and Żuk, 2022
Restorative Justice	Repairing past and ongoing harm. In the context of energy ageism, this means correcting the harm caused by decades of policy neglect that has left many seniors in cold, inefficient homes. It calls for proactive measures to remediate these conditions, not just prevent future harm.	What proactive measures can be taken to upgrade the homes of seniors who have long suffered from energy poverty? How can we compensate for the health impacts caused by living in poorly heated homes or polluted environments?	Jenkins et al. 2016

Table 6.5 synthesizes the concepts of energy justice and energy ageism by demonstrating how each core tenet of energy justice—distributive, procedural, recognition, and restorative—can be used as a lens to analyze and address the specific injustices faced by older people in the low-carbon energy transition. For each framework, the table explains its direct application to energy ageism, such as the unfair distribution of costs or the exclusion from decision-making, and poses key questions that policymakers must consider to create a more equitable, age-inclusive modern energy system. By mapping the principles of energy justice onto the specific challenges of energy ageism and providing main academic references, the table offers a robust conceptual tool for understanding and combating age-based energy inequalities.

The theoretical integration of energy ageism and energy justice also has important implications for understanding the relationship between individual and structural factors in shaping older adults' energy experiences (McCauley et al., 2013). While individual factors such as income, health status, and housing characteristics are important determinants of energy access and participation, the energy ageism framework emphasizes that these individual factors operate within broader structural contexts that can either amplify or mitigate their effects. This structural understanding suggests that effective responses to energy ageism require both individual-level interventions and broader structural changes that address the institutional and policy factors that create and maintain age-based energy disadvantages.

6.2.3 Contributions to Aging Studies and Energy Research

6.2.3.1 Theoretical Contributions to Aging Studies and Energy Research

The development of energy ageism as a theoretical framework also makes important contributions to both aging studies and energy research by highlighting previously underexamined connections between these fields and providing new insights into how aging intersects with contemporary social and environmental challenges (Calasanti, 2015). For aging studies, the energy ageism framework provides a new lens for understanding how social institutions and policies affect older adults' quality of life, independence, and social participation in ways that have received limited attention from aging scholars but have important implications for understanding successful aging in contemporary societies.

The energy ageism framework contributes to aging studies by highlighting how energy systems and energy policy affect older adults' ability to age in a safe place, maintain independence, and participate in community life in ways that extend beyond traditional concerns about healthcare, income security, and social services (Rowe and Kahn, 1997). Energy access and energy system characteristics affect older adults' comfort, safety, health, and social participation in fundamental ways that have important implications for aging policy and practice but have received limited attention from aging researchers and practitioners.

The integrated energy justice and energy ageism framework also contributes to critical aging studies by providing a structural analysis of age-based disadvantage that moves beyond individualistic approaches to aging that focus primarily on personal characteristics or capabilities (Martinson and Berridge, 2015). The energy ageism framework emphasizes how social, economic, and political factors shape older adults' energy experiences in ways that can be modified through appropriate policy interventions and institutional changes, providing a foundation for more effective advocacy and policy development.

For energy research, the energy ageism framework provides new insights into how energy systems affect different populations and how low-carbon energy transition can be designed to promote rather than undermine social equity and inclusion (McCauley et al., 2019). The framework highlights the importance of considering demographic factors in energy system planning and policy development while also providing tools for analyzing and addressing the social dimensions of low-carbon energy transition, such as energy equity, that are crucial for the effectiveness and sustainability of energy systems transformations.

The energy ageism framework also contributes to energy research by highlighting the importance of user-centered approaches to energy system design and policy development that consider the diverse needs, preferences, and capabilities of different user groups (Bouzarovski et al., 2023). This user-centered perspective is particularly important as energy systems become increasingly complex and digitized, requiring greater attention to how different populations can access and benefit from new energy technologies and services.

6.2.3.2 *Methodological Implications and Research Approaches*

The theoretical development of energy ageism also has important methodological implications for research on aging and energy, suggesting the need for new research approaches that can address the complex and multidimensional nature of age-related energy disadvantage (Jenkins et al., 2016). Traditional research approaches that focus on single dimensions of energy access or that treat age as a simple demographic variable may not be adequate for understanding the complex mechanisms through which energy ageism operates or for developing effective interventions to address it.

The multidimensional nature of energy ageism suggests the need for research approaches that can address economic, technological, spatial, informational, and cultural dimensions simultaneously while also considering how these dimensions interact with other forms of inequality and disadvantage like gender, health, race, and ethnicity, among others (Żuk and Żuk, 2022). This requires interdisciplinary research approaches that bring together insights from aging studies, energy research, environmental science, public policy, and other relevant fields, as well as methodological innovations that can address the complex interactions between different factors and scales of analysis.

The intersectional nature of energy ageism also suggests the need for research approaches that can address multiple forms of inequality simultaneously rather than treating age as an isolated factor (Krekula et al., 2018; Stojilovska et al., 2024). This requires attention to how age intersects with income, race, gender, geography, health status, and other factors in ways that create complex patterns of advantage and disadvantage that cannot be understood by examining any single factor in isolation.

The temporal dimensions of energy ageism also suggest the need for longitudinal research approaches that can track changes in older adults' energy experiences over time and identify the factors that contribute to successful adaptation or continued disadvantage (Upham et al., 2022). Longitudinal research is particularly important for understanding how energy ageism develops and changes over time and for evaluating the long-term effectiveness of interventions designed to promote age inclusion in the low-carbon energy transition.

6.3 Empirical Evidence: Comparative Analysis and International Case Studies

6.3.1 Comparative Welfare State Approaches

One of the most significant empirical contributions to understanding age-just low-carbon transitions comes from comprehensive comparative research examining how different welfare state models affect older adults' experiences of energy transitions across multiple dimensions of access, participation, and benefit distribution. The comparative study by Szulecki et al. (2024) analyzes the senior citizens' perspectives on energy transitions in Poland and Norway and provides crucial insights into how institutional contexts shape the manifestation and impact of energy ageism across different national settings, while also highlighting the complex interactions between welfare state institutions, energy system characteristics, and older adults' experiences of low-carbon energy transitions. This study is particularly valuable because it examines two European countries having significantly different welfare state models, energy system characteristics, and historical experiences with energy sector and energy policy developments, providing important insights into how different institutional arrangements can promote or hinder age-inclusive low-carbon energy transitions. Norway represents a social democratic welfare state model with robust social protections, high levels of public investment in energy infrastructure, and significant oil and gas resources that have funded comprehensive social programs over several decades, creating a context of relatively high economic security and extensive public services that might be expected to mitigate some of the barriers associated with energy ageism.

Also, the Norwegian energy system is characterized by extensive hydroelectric power generation, district heating systems in urban areas, and relatively high levels of energy security that have provided a stable foundation for energy transition planning while also creating specific challenges and opportunities for older adult participation (Szulecki et al., 2024). The abundance of renewable energy resources in Norway has enabled relatively low energy costs for all consumers while also providing a foundation for ambitious climate policies and low-carbon energy transition goals that might create both opportunities and challenges for older adults.

Poland, in contrast, represents a post-socialist transition economy with a more limited welfare state, greater reliance on coal-based energy systems, and more recent experience with market-oriented energy reforms that have created different challenges and opportunities for the low-carbon energy transition of the older population (Szulecki et al., 2024). The Polish energy system has historically been dominated by coal-fired power generation, creating significant air quality challenges while also providing relatively affordable energy for consumers. However, the recent European Union climate change mitigation policies and energy market reforms have created pressures for rapid low-carbon energy transition that may create particular challenges for older adults who are dependent on traditional energy sources and may have limited resources for adapting to the new energy systems (Karpinska and Smiech, 2023; Torrego-Gomez et al., 2024).

The empirical findings from this comparative study reveal several important insights about the relationship between welfare state institutions and energy ageism that have broader implications for understanding how different policy approaches can mitigate or exacerbate age-related barriers to energy participation across different national and regional contexts. Most significantly, the research demonstrates that socioeconomic conditions outweigh ageism; therefore, the resilience of the older population in dealing with energy poverty during a low-carbon energy transition is conditioned by their material standing and robustness of the welfare state rather than their age. (Szulecki et al., 2024). This finding suggests that while age-based barriers to energy participation are real and significant, they are often mediated by broader socioeconomic factors and institutional contexts that can be modified through appropriate policy interventions and institutional reforms.

In the Norwegian context, the study found that robust welfare state institutions and higher levels of economic security among older adults helped mitigate some of the barriers associated with low-carbon energy transitions, creating a context in which older adults were generally better positioned to navigate energy system changes and access available programs and funding sources. Norwegian seniors generally had greater financial resources to invest in energy efficiency improvements or new energy technologies, better access to information and support services, and more confidence in their ability to navigate energy system changes due to higher levels of trust in public institutions and greater experience with successful social policy interventions in Norway (Szulecki et al., 2024).

The comprehensive social safety net in Norway also meant that older adults were less vulnerable to energy poverty and had more options for addressing energy-related challenges when they occur, including access to energy efficiency programs, healthcare services, and other social supports that could help mitigate the negative impacts of energy system changes on older adults. The Norwegian welfare state model also includes more extensive public investment in energy infrastructure, which helps to ensure that the benefits of clean energy transitions are more broadly distributed across the entire population. However, even in the Norwegian context, the study identified significant challenges related to digital competence and technological barriers that suggest that robust welfare states may not automatically address all dimensions of energy ageism, particularly those related to the increasing digitization of energy systems and the growing complexity of energy technologies (Szulecki et al., 2024). Many Norwegian seniors expressed concerns about their ability to navigate increasingly digitized energy systems, including smart meters, online energy-management platforms, and digital customer service interfaces that require specific digital skills and confidence that not all older adults possess.

These findings suggest that technological change in energy systems may create new forms of exclusion that require targeted interventions even in countries with comprehensive social protection systems, highlighting the importance of combining comprehensive social protection with targeted interventions that address specific age-related barriers to participation in low-carbon energy transition processes. The Norwegian case demonstrates that while robust welfare states can provide important foundations for age-inclusive low-carbon energy transitions, they are not sufficient by themselves to address all important dimensions of energy ageism.

In the Polish context, the study found that older adults faced more significant barriers to participating in low-carbon energy transitions, reflecting both lower levels of economic security and less comprehensive social support systems that left older adults more vulnerable to energy poverty and less able to access the benefits of low-carbon energy transitions (Szulecki et al., 2024). Polish seniors were more likely to experience energy inequity and poverty, had fewer resources to invest in energy efficiency improvements, and faced greater challenges accessing information and support services related to low-carbon energy transitions, creating a context in which energy ageism was more pronounced and had more severe impacts on older adults' quality of life and their well-being. The study also found that Polish seniors were more dependent on traditional heating sources, particularly coal, which created additional vulnerabilities during periods of energy price volatility and policy change while also creating health risks from air pollution that disproportionately affected older adults with respiratory or cardiovascular conditions. The dependence on coal in heating also created barriers to participating in low-carbon energy transitions, as replacing coal in heating systems with cleaner alternatives requires significant upfront investments and technical knowledge that may not be accessible to older adults with limited economic resources and social support.

The comparative analysis also revealed important differences in how older adults in the two countries perceived and responded to low-carbon energy transition policies, with implications for understanding how trust in institutions and previous experience with policy change affect older adults' willingness to participate in low-carbon energy transitions and support ambitious climate change mitigation policies. Norwegian seniors generally expressed more confidence in government energy and climate policies and more willingness to participate in low-carbon energy transition programs, reflecting higher levels of trust in public institutions and greater experience with successful social policy interventions that had improved their quality of life over time (Szulecki et al., 2024).

Polish seniors, in contrast, expressed more skepticism about low-carbon energy transition policies and greater concerns about the potential costs and disruptions associated with transformations of energy systems, reflecting lower levels of trust in public institutions and more negative experiences with previous energy and social policy changes that had created economic hardship or social disruption. This skepticism created additional barriers to low-carbon energy transition participation and suggested the need for different approaches to communication and engagement that could address older adults' concerns while building trust and confidence in low-carbon energy transition policies.

6.3.2 Social and Spatial Determinants of Age-related Inequities

Additional empirical research has examined the spatial and social determinants of energy ageism in greater detail, providing comprehensive insights into how built environment factors, housing characteristics, and geographic location affect older adults' experiences of low-carbon energy transitions in ways that create complex patterns of advantage and disadvantage that vary significantly across different spatial scales and contexts. The detailed research by Żuk and Żuk (2024a) on social and spatial determinants of energy ageism in Polish society provides particularly valuable insights into these relationships through comprehensive qualitative analysis of older adults' experiences across different housing types and diverse geographic locations.

This research employed focus group interviews with eight groups of senior citizens having different places of residence, types of buildings inhabited, and sources of home heating, providing a comprehensive view of how spatial factors interact with age-related energy vulnerabilities to create different patterns of energy access and participation across different contexts. The spatial focus of this research is particularly important because it recognizes that energy ageism is not uniformly distributed across geographic areas or housing types but rather varies significantly based on local conditions and built environment characteristics that reflect historical patterns of development and investment that have created uneven geographies of opportunity and disadvantage for older adults.

The empirical findings reveal that the type of building and its location within urban space serve as significant determinants of energy poverty among older adults, with implications that extend beyond individual households to encompass neighborhood-level effects and broader patterns of urban inequality that concentrate energy disadvantage in certain areas and housing types. According to (Żuk and Żuk, 2024a), the type of building and its place in the urban space may be a determinant of the energy poverty of its inhabitants, and the area where specific types of buildings are clustered together may affect the air quality of the living environment. This finding highlights how spatial patterns of energy disadvantage can create cumulative effects that affect entire communities and neighborhoods, creating additional challenges for addressing energy ageism that require coordinated interventions addressing its multiple scales simultaneously.

Older adults living in different types of buildings—from historic residences to socialist-era apartment blocks to single-family homes—face different challenges and opportunities related to the implementation of new energy technologies, energy efficiency, heating costs, and access to modern energy services that reflect both the physical characteristics of their housing and the broader social and economic contexts in which they live (Żuk and Żuk, 2024a). The study found that older adults living in older housing stock, particularly in buildings constructed before modern energy efficiency standards were implemented, face significant challenges related to high heating costs and energy poverty risks that are compounded by limited opportunities for improvement due to building ownership structures, financing constraints, and many technical challenges.

These old buildings often have poor insulation, inefficient heating systems, and limited opportunities for energy efficiency improvements that could reduce energy costs and improve comfort levels, creating a situation where older adults living in these buildings face systematically higher energy costs and lower levels of comfort and health compared to those living in newer, more energy-efficient housing. The costs of retrofitting these buildings for improved energy efficiency are often prohibitive for older adults due to their fixed low incomes. This creates a situation where those who most need energy efficiency improvements are least able to afford them, while building owners may lack incentives to invest in improvements that primarily benefit tenants rather than property values. This creates a form of spatial inequality in areas where older adults are living in certain types of housing. The spatial concentration of older, less energy-efficient housing in certain neighborhoods also creates additional challenges, as these areas may have limited access to energy services, higher levels of air pollution, and fewer resources for addressing energy-related problems.

Geographic location within urban areas also emerged as a significant factor affecting older adults' energy experiences, with research revealing that certain neighborhoods and districts have higher concentrations of older, less energy-efficient buildings that create spatial clusters of energy poverty with cumulative effects that extend beyond individual buildings to affect entire communities (Żuk and Żuk, 2024a). These areas may also have poorer air quality due to the concentration of inefficient heating systems and may have limited access to newer energy infrastructure or services, creating additional barriers to energy access and participation in low-carbon energy transition that require coordinated interventions addressing multiple factors simultaneously.

The spatial clustering of energy disadvantage creates additional challenges for addressing energy ageism because it requires coordinated interventions that address neighborhood-level factors rather than simply individual household characteristics, while also creating opportunities for targeted interventions that can reach large numbers of older adults efficiently through area-based approaches. The conducted study suggests that effective responses to spatial dimensions of energy ageism and energy poverty require understanding and addressing the historical patterns of urban development and investment that have created these spatial inequalities while also developing new approaches to energy planning and investment that can address neighborhood-level factors.

The source of home heating emerged as a particularly important factor in determining older adults' vulnerability to energy poverty and their ability to participate in low-carbon energy transitions, with different heating sources creating different patterns of advantage and disadvantage that reflect both technical characteristics and broader social and economic factors (Żuk and Żuk, 2024a). The study found that older adults using different heating sources—including coal, natural gas, district heating, and electric heating—face different challenges and opportunities related to energy costs, environmental impacts, and technological change that require different types of policy responses and support mechanisms.

Older adults who rely on coal heating, which remains common in many parts of Poland and other post-socialist countries, face particular challenges related to both environmental impacts and economic vulnerability that create multiple forms of disadvantage simultaneously. Coal heating requires significant physical labor for fuel procurement and system maintenance, which may become more difficult as individuals age and experience changes in physical capacity, while also creating health risks from indoor air pollution that may be particularly concerning for older adults with respiratory or cardiovascular conditions that are more common in older age groups (Żuk and Żuk, 2024a).

Coal heating also produces significant outdoor air pollution, creating health risks that affect entire neighborhoods and communities while contributing to broader environmental problems that may disproportionately affect older adults who spend more time in their homes and neighborhoods. Additionally, coal heating systems may be difficult or expensive to replace with cleaner alternatives, creating barriers to participating in low-carbon energy transitions that require substantial upfront investments and technical knowledge that may not be accessible to older adults with limited economic resources or skills training support. The research found that older adults using coal heating often face difficult choices between maintaining their current heating systems and investing in cleaner alternatives that may be more expensive in the short term but could provide long-term benefits in terms of both cost savings and health improvements. These choices are complicated by uncertainty about future energy prices, policy changes, and technology developments, as well as concerns about the reliability and performance of new heating technologies that older adults may not be familiar with.

The study showed that older adults using district heating systems face different challenges, including limited control over heating costs and temperatures, dependence on obsolete district heat supply infrastructure that may be unreliable, and limited opportunities to improve energy efficiency through individual actions (Żuk and Żuk, 2024a). However, district heating users also face fewer barriers to participating in low-carbon energy transitions, as system-wide improvements can benefit all users without requiring individual investments in new technologies, suggesting that collective approaches to energy system improvement may be particularly beneficial for older adults who may have limited resources for individual investments.

Table 6.6 presents the spatial and social determinants of energy ageism.

Table 6.6 describes the primary drivers of energy ageism, dividing them into two interconnected types: spatial and social determinants. The spatial determinants focus on the physical environment, detailing how factors like building type, geographic location (urban vs. rural), and access to modern energy infrastructure create the conditions for energy poverty among seniors. The social determinants section outlines the human and systemic factors, explaining

Table 6.6 Spatial and social determinants of energy ageism. Source: created by the author based on Walker (2012), and Żuk and Żuk (2024a,b).

Type	*Determinant*	*Description*	*Impact on energy ageism*
Spatial determinant	Building type and age	The physical characteristics of housing, such as old tenement houses with high ceilings and poor insulation, postwar blocks of flats, or rural single-family homes.	Older buildings are less energy-efficient, leading to higher heating costs and greater energy poverty. Seniors living in them are often unable to afford necessary thermal comfort.
	Geographic location (urban vs. rural)	The location of a residence in either a dense urban center, a neglected city district, or a rural area.	Rural areas often lack access to district heating or gas networks, forcing reliance on solid fuels. Neglected urban districts face similar infrastructure gaps and are often home to low-income seniors.
	Infrastructure access	The availability of modern energy infrastructure, such as a municipal district heating network, a gas network, or a stable electrical grid capable of supporting new technologies.	Lack of access to efficient infrastructure traps seniors in expensive, polluting, and labor-intensive heating methods (e.g., coal or wood), directly contributing to energy poverty and poor health.
	Neighborhood characteristics	The socioeconomic status of a district, including whether it is gentrifying, neglected, or suburbanizing. This also includes poor local air quality.	Seniors in neglected districts suffer from poor air quality caused by neighbors burning low-quality fuel. In gentrifying areas, they may be pushed out by rising costs, while suburban areas can create isolation.

Table 6.6 (continued)

Type	*Determinant*	*Description*	*Impact on energy ageism*
Social determinant	Socioeconomic status (income and class)	The financial resources of older people, primarily determined by low pensions, lack of employment opportunities, and overall wealth.	This is the most direct determinant. Low, fixed incomes make it impossible for seniors to absorb rising energy prices, pay for furnace replacements, or make the co-payments required for energy efficiency or retrofitting programs.
	Household composition	Whether a senior lives alone (especially widowed individuals), with a partner, or in a multi-generational household.	Single-person households are far more vulnerable to energy poverty, as they have a single, often low, fixed income to cover all fixed costs. In contrast, intergenerational solidarity can provide financial and technical support for older adults.
	Policy and institutional failures	The design of social and energy policies, including the welfare system, subsidy programs for low-carbon energy transition, and the level of support from local authorities.	Bureaucratic, inaccessible, and underfunded energy efficiency programs with requirements like digital literacy or upfront costs effectively exclude the most vulnerable seniors, reinforcing the "vicious cycle" of energy ageism.
	Cultural and educational factors	An individual's level of trust in authorities and new technologies, their access to information, and their understanding of the low-carbon energy transition.	A lack of accessible information and distrust stemming from past negative experiences can lead to skepticism and resistance to adopting new, more efficient technologies, even when financial support is available for older adults.

how socioeconomic status, household composition, institutional policy failures, and cultural barriers directly impact an older person's ability to cope with the low-carbon energy transition and to access its benefits. By presenting these determinants side-by-side, the table illustrates the core argument: that energy ageism is not simply a matter of individual hardship but a complex issue rooted in the interplay between where older people live and the social structures that govern their lives.

The spatial and social determinants research demonstrates that energy ageism operates through complex interactions between individual, household, community, and structural factors that create uneven geographies of energy disadvantage and opportunity. This research highlights the importance of understanding how built environment factors, housing characteristics, and geographic location interact with age-related vulnerabilities to create complex patterns of energy disadvantage that require comprehensive policy responses addressing multiple scales and dimensions simultaneously.

6.3.3 Digital Competence, Economic Vulnerability, and Health Impacts

6.3.3.1 Digital Competence and Energy Transition Participation

A consistent finding across multiple empirical studies is the significance of digital competence as a factor affecting older adults' ability to participate effectively in low-carbon energy transitions, with implications that extend beyond simple access to energy technology to encompass broader questions of social inclusion and participation in modern energy systems that are becoming increasingly digitized and automated (Bouzarovski et al., 2023). The increasing digitization of energy systems—including smart meters, use of AI, online energy-management platforms, digital customer service interfaces, and mobile applications for energy monitoring—creates new requirements for digital skills and access that may pose challenges for some older adults while providing opportunities for others to optimize their energy consumption and reduce their energy costs.

The empirical studies showed that complex relationships between age, digital competence, and energy system participation that resist simple generalizations about older adults' technological capabilities highlight the importance of understanding the diverse factors that affect digital access and skills among older adults (Szulecki et al., 2024; Bouzarovski et al., 2023). While some older adults demonstrate high levels of digital competence and enthusiasm for new energy technologies, others face significant barriers related to digital access, skills, or confidence that can limit their ability to participate fully in increasingly digitized energy systems, creating new forms of inequality that may compound existing age-related disadvantages.

These differences in digital competence and access are influenced by factors such as education, income, health status, social support networks, and previous experience with energy technology, suggesting that digital barriers to energy participation are not simply a function of age but reflect broader patterns of social inequality that intersect with age in complex ways (Furszyfer Del Rio et al., 2021). The research by Szulecki et al. (2024) found that digital competence emerged as a key factor distinguishing older adults who were able to participate effectively in low-carbon energy transitions from those who faced barriers to participation, with implications for both individual outcomes and broader energy system effectiveness.

Older adults with higher levels of digital competence were more likely to access information about energy programs and services, navigate online energy-management platforms, and take advantage of digital tools for energy monitoring and optimization that could provide for them both environmental and economic benefits (Yagita and Iwafune, 2021). These older adults were also more likely to participate in energy efficiency and energy retrofit programs and other community-energy initiatives and to engage with energy policy discussions through online platforms and digital communication channels, suggesting that digital competence affects not only individual energy outcomes but also broader patterns of civic participation and democratic engagement.

However, the empirical studies also found that digital barriers were not unbeatable and that appropriate support and education programs could help older adults develop their digital skills needed to participate in low-carbon energy transitions while also building confidence and enthusiasm for new energy technologies. According to Szulecki et al. (2024), it is most important to mainstream the social inclusion considerations in energy policy and to build targeted digital competence, which can enhance senior citizen's effective integration in the low-carbon energy transition. This finding suggests that addressing digital barriers requires both individual-level interventions such as digital-literacy training programs and system-level changes such as ensuring that energy services remain accessible through non-digital channels.

The research on artificial intelligence and energy justice has also revealed important concerns about how algorithmic decision-making in energy systems may create new forms of discrimination that disproportionately affect older adults and other vulnerable populations (Furszyfer Del Rio et al., 2021). As energy systems become increasingly automated and rely on artificial intelligence for functions such as demand forecasting, price setting, and service delivery, there is growing concern that these systems may disadvantage older adults or other vulnerable populations through automated decision-making processes that may not adequately consider their specific needs and circumstances.

These algorithmic biases may operate through multiple mechanisms, including training data that does not adequately represent older adult populations, algorithmic models that prioritize efficiency over equity, and automated decision-making processes that do not include appropriate safeguards against age and other discrimination. The study (Furszyfer Del Rio et al., 2021) suggests that addressing these concerns requires not only technical improvements to algorithmic systems but also broader changes in how energy systems are governed and regulated to ensure that automated decision-making processes are transparent, accountable, and fair.

The findings from empirical case studies linked to age-just, low-carbon energy transitions suggest that successful approaches to digital inclusion for older adults in low-carbon energy transitions require understanding and addressing the diverse factors that affect digital competence and access, including not only individual skills and resources but also the design of digital systems in energy usage and the availability of alternative access channels. The research by Bouzarovski et al. (2023) indicates that effective digital inclusion strategies must address both the supply side of digital services through improved design and accessibility and the demand side through education, support, and capacity-building programs that are tailored to older adults' specific needs and preferences.

6.3.3.2 Economic Vulnerability and Energy Poverty

Empirical research consistently identifies economic vulnerability as a central factor affecting older adults' negative experiences of low-carbon energy transitions. Energy poverty—defined as the inability to afford adequate energy services for basic needs—disproportionately affects older adults due to a combination of factors, including fixed low incomes, higher energy consumption patterns, and health conditions that require additional energy use for heating, cooling, or medical equipment (Żuk and Żuk, 2024a, b). The study shows that energy poverty among older adults is not simply a matter of absolute income levels but reflects complex interactions between income, housing characteristics, energy prices, and energy system design that create different patterns of vulnerability across different groups of older adults.

The comparative research mentioned above analyzing differences in low-carbon transition experiences of older adults in Poland and Norway demonstrates how different economic contexts and social policy approaches affect older adults' vulnerability to energy poverty, with implications for understanding how welfare state institutions can mitigate or exacerbate energy-related disadvantages (Szulecki et al., 2024). In Norway, comprehensive social programs and higher pension levels help protect older adults from energy poverty, while in Poland, lower income levels and less comprehensive social support create greater vulnerability to energy cost increases and energy system changes that may disrupt traditional energy sources and practices.

The study highlights that energy poverty among older adults has implications beyond individual hardship, affecting their ability to participate in low-carbon energy transitions and access the benefits of clean energy technologies that could provide long-term economic and environmental benefits. Older adults experiencing energy poverty may be unable to invest in energy efficiency improvements or renewable energy technologies that could reduce their long-term energy costs, creating a situation where those who most need energy cost reductions are least able to access them (Żuk and Żuk, 2024a, b). This creates a form of economic exclusion from low-carbon energy transitions that can perpetuate and exacerbate existing inequalities while limiting the effectiveness of energy transition policies that rely on voluntary adoption of new renewable energy technologies and practices.

This economic exclusion from low-carbon energy transitions creates additional challenges for achieving ambitious climate change mitigation goals, as it limits the potential for demand-side energy savings and renewable energy adoption that depend on broad participation across all segments of the population. The study (Żuk and Żuk, 2024a) suggests that addressing economic barriers to low-carbon energy transition participation requires not only income-support programs but also targeted interventions that help older adults access the upfront capital needed for energy investments and navigate complex financing arrangements that may be difficult to understand or access without appropriate support.

The spatial dimensions of economic vulnerability also emerge as important factors as older adults usually living in certain types of housing and geographic areas are facing systematically higher energy costs and fewer opportunities for energy savings (Żuk and Żuk, 2024a, b). These spatial patterns of economic vulnerability reflect historical patterns of investment and disinvestment that have created uneven geographies of energy access and opportunity that intersect with age-related vulnerabilities in complex ways, requiring coordinated interventions that address both individual economic circumstances and broader spatial inequalities.

Therefore, the economic vulnerability among older adults is often compounded by other factors such as health conditions that require additional energy use, social isolation that limits access to information and support, and housing characteristics that create higher energy costs or limit opportunities for energy efficiency improvements. These multiple forms of vulnerability create cumulative disadvantages that require comprehensive interventions addressing multiple factors simultaneously rather than focusing on any single dimension of economic disadvantage.

6.3.3.3 Health and Well-being Impacts

Empirical research has also documented significant relationships between energy systems, low-carbon energy transitions, and older adults' health and well-being that operate through multiple pathways and have important implications for understanding the full scope of energy ageism and its impacts on older adults' quality of life (Żuk and Żuk, 2024a, b). These relationships include direct health impacts from energy-related air pollution, indirect health impacts from energy poverty and inadequate heating or cooling of homes, and psychosocial impacts from energy system changes and energy insecurity that can affect older adults' mental health and social well-being in ways that extend beyond immediate physical health concerns.

The research on spatial determinants of energy ageism found that older adults living in areas with high concentrations of inefficient heating systems face greater exposure to air pollution, which can exacerbate respiratory and cardiovascular conditions that are more common among older adults and may have more severe health impacts due to age-related changes in immune function and physiological resilience (Żuk and Żuk, 2024a). The transition to cleaner energy sources has the potential to provide significant health benefits for older adults, but only if the transition is implemented in ways that ensure equitable access to clean energy technologies and services while also addressing the spatial concentrations of pollution that disproportionately affect certain communities and neighborhoods.

Energy poverty also has direct health impacts for older adults, who may be particularly vulnerable to temperature extremes and may have health conditions that require consistent heating or cooling for medical reasons, such as respiratory conditions that are exacerbated by cold temperatures or cardiovascular conditions that are affected by heat stress (Żuk and Żuk, 2024a). The older adults experiencing energy poverty may face difficult choices between paying for energy and paying for other necessities such as food or medication, creating additional health risks that extend beyond the direct effects of inadequate heating or cooling to encompass broader impacts on nutrition, medication adherence, and healthcare access.

The psychosocial impacts of energy system changes also emerge as significant factors affecting older adults' well-being, with energy transitions potentially creating uncertainty and anxiety for older adults who are concerned about their ability to navigate new energy systems or afford new energy technologies (Szulecki et al., 2024). These psychosocial impacts may be particularly significant for older adults who have limited social support networks or who have experienced previous disruptions to their living situations, creating additional stress and anxiety that can affect both mental and physical health outcomes.

Therefore, the energy insecurity—uncertainty about the ability to afford adequate energy services—can have significant impacts on older adults' mental health and social well-being, creating chronic stress and anxiety that can affect sleep, appetite, social participation, and overall quality of life. Energy insecurity may also affect older adults' relationships with family members and social networks, as concerns about energy costs may limit their ability to host social gatherings or participate in community activities that require energy use.

6.3.3.4 Community and Social Capital Factors

Empirical research has also identified community and social capital factors as important determinants of older adults' experiences of low-carbon energy transitions. The older adults who have strong social support networks, active community connections, and access to informal sources of information and assistance are better positioned to navigate energy system changes and access available resources and programs (Szulecki et al., 2024). Therefore, it is important to understand that low-carbon energy transitions are not simply technical or economic processes but social processes that are shaped by community relationships and social networks that can either facilitate or hinder older adults' participation in energy system changes and obtaining benefits from low-carbon energy transitions.

The community-based approaches to low-carbon energy transitions can be particularly effective for engaging older adults and addressing their specific needs and concerns, with community energy programs that involve older adults in energy efficiency and renewable energy technology implementation planning, provide peer-to-peer learning opportunities, and offer ongoing support and assistance showing promise for promoting more inclusive low-carbon energy transitions (McCauley et al., 2013). However, many older adults may be excluded from community-energy initiatives due to barriers such as meeting formats that are not accessible, communication strategies that do not reach older adult communities, or decision-making processes that do not accommodate diverse participation styles and older adult preferences (McCauley et al., 2013).

Addressing these barriers requires intentional efforts to design community-energy programs that are inclusive of older adults and responsive to their needs and preferences, including consideration of factors such as meeting times and locations, communication methods, and decision-making processes that accommodate different levels of technical knowledge and different participation preferences (Fuller and McCauley, 2016). Therefore, the successful community-energy programs for older adults often involve partnerships with existing community organizations and service providers that already have established relationships and trust with older adult populations and communities.

The role of social capital in mediating older adults' experiences of low-carbon energy transitions also highlights the importance of addressing social isolation and building community connections as part of comprehensive approaches to age-just, low-carbon energy transitions. Older adults who are socially isolated may face additional barriers to accessing necessary information, support, and resources related to low-carbon energy transitions, while those with strong community connections may be better positioned to benefit from energy system changes and to contribute to community-energy initiatives (Szulecki et al., 2024).

The peer networks and informal information sharing play important roles in how older adults learn about and respond to low-carbon energy transition opportunities, with older adults often relying on recommendations from friends, neighbors, and family members when making decisions about energy investments or program participation. This suggests that effective outreach and education strategies for older adults should include peer-to-peer components and should work through existing social networks and older adults' community organizations rather than relying solely on formal communication channels.

The empirical evidence reviewed demonstrated that energy ageism is not merely a theoretical concern but a real phenomenon that affects older adults' quality of life, independence, and social participation in measurable ways across different national contexts and institutional arrangements. The comparative research from different national contexts reveals that while robust welfare state institutions can help mitigate some age-related barriers to energy participation, targeted interventions are necessary to address the specific challenges that older adults face in the context of energy transitions, particularly those related to digital competence, technological change, and cultural recognition.

6.4 International Comparative Insights

Beyond the detailed Poland-Norway comparison, emerging research from other national contexts provides additional insights into how different institutional arrangements, cultural contexts, and energy system characteristics affect older adults' experiences of low-carbon energy transitions (Upham et al., 2022). These international comparative perspectives are important for understanding the generalizability of findings from specific national contexts and for identifying best practices and successful models that can be adapted to different settings while also highlighting the importance of context-specific approaches to addressing energy ageism.

Research from other European countries with different welfare state models and energy system characteristics provides additional evidence about the relationships between institutional contexts and energy ageism, while also highlighting the diversity of approaches that different countries have taken to addressing age-related energy concerns (Heffron, 2022). Countries with different energy resource endowments, different historical experiences with energy policy development, and different demographic profiles may face different challenges and opportunities for promoting age-inclusive low-carbon energy transitions.

The cultural factors and social attitudes toward aging may play important roles in shaping how low-carbon energy transitions affect older adults, with countries that have more positive attitudes toward aging and stronger traditions of intergenerational solidarity potentially being better positioned to develop age-inclusive energy policies and programs (Ayalon and Tesch-Römer, 2018). However, according to a study by Ayalon and Tesch-Römer (2018), the cultural factors alone are not sufficient to address energy ageism, and institutional and policy interventions are necessary to ensure that positive attitudes toward aging are translated into concrete benefits for older adults.

Studies also highlight the importance of learning from different approaches to low-carbon energy transition governance and policy development, with some countries developing more participatory and inclusive approaches to energy planning that may be better positioned to address age-related concerns (Jenkins et al., 2016). According to Jenkins et al. (2016), the countries moving toward more participatory, inclusive energy-planning processes and addressing group-specific needs (including age-related ones) are Germany, Denmark, Netherlands, France, and Ireland. These comparative insights suggest that promoting age-just, low-carbon energy transitions requires not only technical and economic innovations but also governance and policy innovations that can address the complex social and political dimensions of energy system transformation.

References

Agrawal, K.; Pathak, M.; Jana, K.; Unni, J.; Shukla, P. Just transition away from coal: vulnerability analysis of coal districts in India, Energy Research & Social Science, 2024, 108(2024), 103355.

Axon, S.; Morrissey, J. Just energy transitions? Social inequities, vulnerabilities and unintended consequences, Build. Cities, 2020, 1(1), pp. 393–411.

Ayalon, L.; Tesch-Römer, C. Contemporary perspectives on ageism. Springer, 2024. https://doi.org/10.1007/978-3-319-73820-8.

Bouzarovski, S.; Fuller, S.; Reames, T.G. Handbook on Energy Justice. Cheltenham, UK: Edward Elgar Publishing, 2024. ISBN: 978 1 83910 295 0.

Bouzarovski, S.; Petrova, S. A global perspective on domestic energy deprivation Overcoming the energy poverty–fuel poverty binary, Energy Research and Social Science, 2020, 10, pp. 31–40.

Butler, R.N. Age-ism: Another form of bigotry. The Gerontologist, 1969, 9(4), pp. 243–246.

Calasanti, T. Combating ageism: How successful is successful aging? The Gerontologist, 2015, 56(6), pp. 1093–1101.

Cha, J.M. A just transition for whom? Politics, contestation, and social identity in the disruption of coal in the Powder River Basin, Energy Research & Social Science, 2020, 69, 101657.

Cebola, M.M.J.; Dos Santos, N.R.; Dionísio, A. Worker-related ageism: a systematic review of empirical research, Ageing & Society, 2023, 43 (8), pp. 1882–1914.

Ciplet, D. From energy privilege to energy justice: A framework for embedded sustainable development. Energy Research and Social Science, 2021, 75, 101996.

Creti, A.; Ftiti, Z. Energy, just transition, and sustainability: What's new? Energy Economics, 2024, 139, 107872.

Eaton, E. Approaches to energy transitions: carbon pricing, managed decline, and/ or green new deal? Geogr. Compass, 2020, 15(2), e12554.

Endres, D. The rhetoric of nuclear colonialism: Rhetorical exclusion of American Indian arguments in the Yucca Mountain nuclear waste siting decision. Communication and Critical/Cultural Studies, 2009, 6(1), pp. 39–60.

Fenger, H.J.M. Welfare Regimes in Central and Eastern Europe: Incorporating Post-Communist Countries in a Welfare Regime Typology. Contemporary Issues and Ideas in Social Sciences. Journal of European Social Policy, 2007, 17(2), pp. 1–30.

Fuller, S.; McCauley, D. Framing energy justice: perspectives from activism and advocacy. Energy Research and Social Science, 2016, 11, pp. 1–8.

Furszyfer Del Rio, D.D; Sovacool, B.K.; Griffiths, S. Culture, energy and climate sustainability, and Smart Home Technologies: A mixed methods comparison of four countries. Energy and Climate Change, 2021, 2, 100035.

Goddard, G.; Farrelly, M.A. Just transition management: Balancing just outcomes with just processes in Australian renewable energy transitions. Applied Energy, 2021, 225, pp. 110–123.

Healy, N., Barry, J. Politicizing energy justice and energy system transitions. Energy Policy, 2017, 107, pp. 451–459.

Heffron, R.J. Applying energy justice into the energy transition. Renewable and Sustainable Energy Reviews, 2022, 156, 111936.

Jenkins, K. Setting energy justice apart from the crowd: Lessons from environmental and climate justice. Energy Research and Social Science, 2018, 39, pp. 117–121.

Jenkins, K.; McCauley, D.; Heffron, R.; Stephan, H.; Rehner, R. Energy justice: A conceptual review. Energy Research and Social Science, 2016, 11, pp. 174–182.

Jenkins, K.; Sovacool, B.K.; McCauley, D. Humanizing sociotechnical transitions through energy justice: An ethical framework for global transformative change. Energy Policy, 2018, 117, pp. 66–74.

Karpinska, L.; Smiech, S. Multiple faces of poverty. Exploring housing-costs-induced energy poverty in Central and Eastern Europe, Energy Research and Social Science, 2023, 105, 103273.

Krekula, C.; Nikander, P.; Wilińska, M. Multiple marginalizations based on age: Gendered ageism and beyond. In L. Ayalon and C. Tesch-Römer (Eds.), Contemporary Perspectives on Ageism (pp. 33–50). Cham: Springer, 2018. https://doi.org/10.1007/978-3-319-73820-8_3.

Levin, L.; Nevo, M. Person, people, planet: eco-systematic analysis of older adults' experiences of engagement with nature and discourse about nature. Journal of Gerontological Social Work, 2024, 67(4), pp. 541–557.

Levy, B.R.; Banaji, M.R. Implicit ageism. In T. D. Nelson (Ed.), Ageism: Stereotyping and Prejudice Against Older Persons (pp. 49–75). 2002, Cambridge, MA: MIT Press. ISBN 978-0-262-14077-0. https://direct.mit.edu/books/edited-volume/2689/chapter/72722/Implicit-Ageism

Martinson, M.; Berridge, C. Successful aging and its discontents: A systematic review of the social gerontology literature. The Gerontologist, 2015, 55(1), pp. 58–69.

Memmott, T.; Carley, S.; Graff, M.; Konisky, D. M. Sociodemographic disparities in energy insecurity among low-income households before and during the COVID-19 pandemic, Nature Energy, 2021, 6(2), pp. 186–193.

McCauley, D.; Heffron, R.; Stephan, H.; Jenkins, K. Advancing Energy Justice: The Triumvirate of Tenets. International Energy Law Review, 2013, 32(3), pp. 107–110

McCauley, D.; Ramasar, V.; Heffron, R.J.; Sovacool, B.K.; Mebratu, D.; Mundaca, L. Energy justice in the transition to low carbon energy systems: Exploring key themes in interdisciplinary research. Applied Energy, 2019, 233–234, pp. 916–921.

Rowe, J. W., Kahn, R. L. (1997). Successful aging. The Gerontologist, 1997, 37(4), pp. 433–440.

Schlosberg, D. Defining Environmental Justice: Theories, Movements, and Nature. 2007, Oxford University Press: New York.

Shrader-Frechette, K. Environmental Justice: Creating Equality, Reclaiming Democracy. 2002. Oxford Universitry Press: New York, p. 284.

Sovacool, B.K. Who are the victims of low-carbon transitions? Towards a political ecology of climate change mitigation, Energy Res. Soc. Sci., 2021, 73(2021), 101916.

Sovacool, B.K.; Burke, M.; Baker, L.; Kotikalapudi, C.K.; Wlokas, H. New frontiers and conceptual frameworks for energy justice. Energy Policy, 2017, 105, pp. 677–691.

Sovacool, B.K.; Dworkin, M.H. Energy justice: Conceptual insights and practical applications. Applied Energy, 2015, 142, pp. 435–444.

Stephens, J.C. Energy democracy: redistributing power to the people through renewable transformation, Environment, 2019, 61(2), pp. 4–13.

Stojilovska, A., Yoon, H., Frankowski, J. Enough pain, cold, and illness! Social movements in response to energy poverty in Europe, Energy Research and Social Science, 2024, 115, 103627.

Szulecki, K.; Neerland, M.A.; Tomter, H.; Wæringsaasen, C.A.B.; Żuk, P.; Żuk, P. Ageism, welfare, and the energy transition: a comparative analysis of the perceptions among the elderly in Poland and Norway. Energy, Sustainability and Society, 2024, 14, p. 35.

Tikkakoski, P.; Aulake, M.; Paloniemi, R. Towards just transition: tackling inequity and structural causes of vulnerability in key environment, health and climate related policies in Finland, Environmental Science & Policy, 2024, 156(2024), 103736.

Torrego-Gomez, D.; Gayoso-Heredia, M.; Núnez-Peiro, M.; Sanchez-Guevara, C. Mapping summer energy poverty: the lived experience of older adults in Madrid, Spain, Energy Research and Social Science, 2024, 110(2024), 103449.

Upham, P.; Sovacool, B.; Ghosh, B. Just transitions for industrial decarbonisation: A framework for innovation, participation, and justice. Renewable and Sustainable Energy Reviews, 2022, 167, 112699.

Walker, A. The new ageism. The Political Quarterly, 2012, 83(4), pp. 812–819. https://doi.org/10.1111/j.1467-923X.2012.02360.x.

Willand, N. Opportunity, ideal or distraction? Exploring stakeholder perceptions of tackling energy poverty and vulnerability among older Australians, Energy Research and Social Science, 2012, 94, 102852.

World Health Organization. World report on ageing and health. World Health Organization. 2015. https://apps.who.int/iris/handle/10665/186463

Yagita, Y.; Iwafune, Y. Residential energy use and energy-saving of older adults: a case from Japan, the fastest-aging country. Energy Research and Social Science, 2021, 75(2021), 102022.

Żuk, P.; Żuk, P. Energy ageism: The framework of the problem and the challenges of a just energy transition. Environmental Innovation and Societal Transitions, 2022, 43, 237–243.

Żuk, P.; Żuk, P. Social and spatial determinants of energy ageism: Calibrating social policy towards older people under the conditions of energy transition in Polish society. Energy Research and Social Science, 2024a, 118, 103795.

Zuk, P.; Zuk, P. Ecology for the rich? Class aspects of the green transition and the threat of right-wing populism as a reaction to its costs in Poland. Sustainability: Science, Practice and Policy, 2024b, 20 (1), 2351231.

Zuk, P.; Zuk, P. National energy security or acceleration of transition? Energy policy after the war in Ukraine, Joule, 2022, 6(4), pp. 709–712.

Żuk, P. The war in Ukraine: consequences for the economy, labour class and equitable development in Europe and beyond. Economic and Labour Labour Relations Review, 2023, 34(2), pp. 343–356.

CHAPTER 7

Policy and Analytical Pathways for Advancing Social Equity in the Energy Sector

7.1 Analytical Approaches to Measuring and Assessing Equity in Energy Systems

7.1.1 Indicators and Metrics of Energy Equity

The assessment of social equity in the energy sector requires systematic indicators and metrics that capture distributive, procedural, and recognition-based dimensions. Indicators function as tools for identifying inequities in access, affordability, and participation, while metrics provide the means to evaluate progress over time and across contexts. Developing robust equity indicators is challenging because equity is inherently multidimensional and shaped by social, economic, and institutional conditions.

Early approaches to equity measurement relied primarily on economic indicators, such as household energy expenditure shares or price-to-income ratios, which provided useful but limited insights into distributive aspects. More recent scholarship has expanded the scope to include social indicators, such as gender, age, and geographic disparities, thereby reflecting the differentiated impacts of energy systems on various population groups (Bartiaux et al., 2018; Tarekegne et al., 2021; Feenstra and Özerol, 2021).

The most frequently applied metrics of social equity in energy assessments can be grouped into several categories (Fig. 7.1):

7.1.1.1 Affordability

Affordability is the cornerstone of equity assessments, and it is often operationalized through economic indicators directly tied to household budgets and market prices, rather than narrow poverty thresholds alone. The most common affordability metrics include the following:

I. Siksnelyte-Butkiene, D. Streimikiene, *Social Equity in the Energy Sector*,
https://doi.org/10.1007/978-3-032-22224-4_7

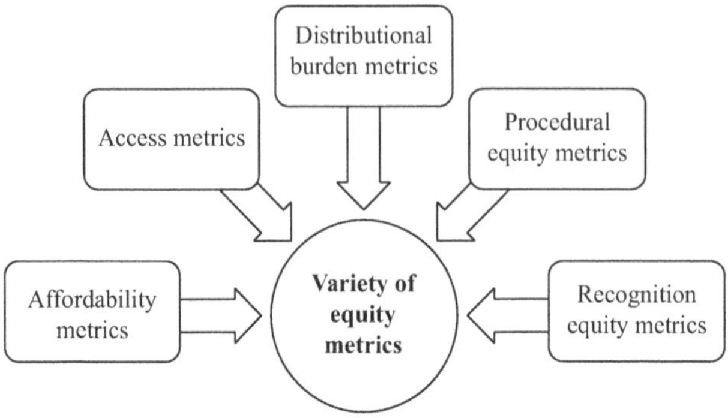

Fig. 7.1 Metrics to measure energy equity. Source: developed by the author.

- Household energy spending (in absolute or relative terms). Household expenditures on electricity, heating fuels, and transport energy are frequently used as direct measures of affordability. These can be expressed in local currency to track national trends or adjusted for purchasing power parity to enable cross-country comparisons. When combined with household survey data, expenditure metrics provide insight into both the absolute financial pressures faced by consumers and relative differences across socioeconomic groups.
- Energy expenditure as a share of household income. This indicator highlights the proportion of household income devoted to energy costs and is widely applied to capture disproportionate burdens. Often referred to as the energy burden, it is particularly effective in revealing inequalities when disaggregated by income or socioeconomic groups. Beyond identifying vulnerable households, this metric is increasingly used in policy evaluation to assess the distributional outcomes of energy price reforms, subsidies, or targeted support schemes.
- Energy price levels and trends. Retail electricity prices, natural gas tariffs, district heating charges, and transport fuel prices are among the most frequently applied affordability indicators. They are systematically reported in recognized international databases, including Eurostat, IEA, and OECD. Tracking price levels and trends is critical for equity analysis, as even moderate price increases can have regressive impacts when low-income households lack the capacity to adjust consumption or switch to alternatives.
- Price-to-income ratios. These indicators compare average residential energy prices—typically for electricity, gas, or heating fuels—with measures of household income such as median disposable income. They provide a standardized way to track affordability across time and space, allowing comparisons between regions with different income levels and price structures. Price-to-income ratios are particularly valuable in international benchmarking and in evaluating the affordability implications of tariff reforms, tax adjustments, or energy price shocks.

- Volatility-adjusted energy prices. Beyond absolute price levels, indicators that capture price variability and inflation dynamics are increasingly important for assessing affordability. Volatility-adjusted metrics consider both the level and the fluctuation of energy costs over time, reflecting households' exposure to sudden or sustained price shocks. This dimension is particularly relevant for low-income households, whose limited financial flexibility amplifies the impact of price instability.
- Arrears on utility bills and disconnections. These indicators serve as direct proxies for affordability stress and are applicable across diverse contexts. Arrears on utility bills capture households' inability to pay for energy on time, while disconnection rates indicate acute vulnerability when households lose access to essential services due to nonpayment. Such indicators are widely monitored by regulators, utilities, and survey programs in many regions. They provide valuable insights into short-term affordability crises, particularly during periods of sharp price increases or economic downturns. Increasingly, arrears and disconnections are used in policy evaluation to assess the effectiveness of targeted transfers, social tariffs, and price caps in preventing energy exclusion and protecting vulnerable consumers.

Many of the affordability indicators outlined above are conceptually similar to the single-indicator approaches that have historically been developed for measuring energy poverty. For instance, the widely used measure of energy expenditure as a share of income directly mirrors the classical '10% rule', which defined households as fuel poor if they devoted more than one-tenth of their income to energy (Boardman, 1991). Although 10% has been the conventional benchmark, other cutoffs—such as 8%, 12%, or 15%—have also been applied in different national contexts, reflecting varying energy needs and income levels (Moore, 2012; Siksnelyte-Butkiene et al., 2021).

The Low-Income High-Cost (LIHC) measure, introduced in the United Kingdom, combines high relative energy costs with low household incomes and has influenced both policy and comparative research (Hills, 2012). A related approach is the After Fuel Cost Poverty (AFCP) measure, which classifies households as energy poor if their residual income, after paying for necessary energy services, drops below the poverty threshold. AFCP is valuable in highlighting "hidden" energy poverty, as it captures households that are not initially poor but fall into poverty once energy expenses are accounted for (Hills, 2011; Moore, 2012).

Other indicators provide relative or outcome-based measures. The median expenditure share (MES) approach, which defines energy poverty when households spend more than twice the national median share of income on energy, provides a relative measure that adapts to national contexts (Heindl and Schüßler, 2015). Similarly, the minimum income standard (MIS) approach identifies households as energy poor if, after paying for energy services, their remaining income falls below a socially defined minimum living standard (Moore, 2012).

In addition to these expenditure-based indicators, outcome-based stress indicators and self-reported measures—such as arrears on utility bills, disconnection rates, and the inability to keep the home adequately warm—have long been used to capture affordability stress and are now integral to both energy poverty and equity assessments (Eurostat, 2025). These measures provide complementary evidence by linking affordability not only to budgets and prices, but also to the lived experiences of households, as only income-based metrics ignore people's experiences and behavior patterns (Cong et al., 2022).

The key distinction lies in the purpose of application. In the energy poverty literature, single indicators were designed primarily to identify vulnerable households within national contexts, often for targeting support programs. In contrast, energy equity assessments employ the same or closely related indicators to evaluate system-wide fairness, examining how the costs of energy supply, taxation, subsidies, or market reforms are distributed across different socioeconomic groups.

This overlap illustrates that affordability indicators serve as a conceptual bridge between energy poverty research and energy equity analysis. Both fields use similar metrics, but with different orientations: energy poverty focuses on identifying deprivation, while equity assessments extend these indicators to assess fairness at the scale of the entire energy system. Recognizing this relationship also highlights the need to move beyond single indicators toward multidimensional and composite frameworks, which can capture not only affordability but also access, procedural fairness, and recognition of vulnerable groups.

A further distinction arises when applying affordability indicators in the Global South and the Global North. In the Global South, affordability assessments often focus on the absolute cost of minimum energy services relative to very low household incomes, with particular attention to expenditures on basic cooking and lighting fuels. Volatility in kerosene, LPG, or electricity prices can push households into energy poverty even when their absolute spending appears low by international standards. Here, indicators such as the cost of minimum energy service baskets, arrears on utility bills, and fuel-switching behaviors provide crucial evidence of vulnerability (International Energy Agency, 2024; IEA et al., 2025).

By contrast, in the Global North, affordability debates are shaped less by the question of basic access and more by relative burdens. Indicators such as energy expenditure shares, the 10% rule, or the LIHC measure dominate assessments, reflecting the concern that rising energy costs or regressive policy instruments (e.g., carbon pricing) disproportionately affect low-income households. Heating costs, electricity tariffs, and energy price volatility are central issues, and disconnection or arrears data are widely used as stress indicators.

These differences highlight the need for context-sensitive measurement. While affordability indicators share common methodological roots, their relevance and interpretation differ across settings: in the Global South they are tied to questions of minimum energy security and poverty alleviation, whereas in the Global North they are primarily used to evaluate the distributional fairness of energy systems and implemented policies.

7.1.1.2 Access Metrics

Equity in energy systems is also fundamentally shaped by the degree to which households and communities have reliable access to modern energy services. Access metrics are widely used in both international monitoring frameworks and academic research to evaluate progress toward universal energy availability. Unlike affordability indicators, which capture the financial dimension of equity, access indicators emphasize the physical availability, quality, and reliability of energy supply. The most common access metrics include:

- Electrification rate. Measured as the share of the population connected to the electricity grid, this indicator has been systematically tracked by the IEA and the World Bank in the context of Sustainable Development Goal 7 (SDG 7). While electrification rates have improved globally, national averages can mask persistent inequalities, particularly in rural areas and among marginalized groups (IEA et al., 2025).
- Access to clean cooking fuels and technologies. This reflects whether households rely on modern cooking solutions (e.g., LPG, electricity, biogas) or continue to depend on traditional biomass. The indicator is especially important because polluting fuels undermine health and disproportionately affect women and children, who are more exposed to indoor air pollution and fuel collection burdens.
- Service quality and reliability. Beyond connection rates, indicators such as average daily hours of electricity supply, frequency and duration of outages, and voltage stability capture whether access translates into effective and usable energy services. These dimensions are incorporated into the World Bank's Multi-Tier Framework (MTF), which grades energy access across tiers of availability, affordability, reliability, legality, and safety (Bhatia and Angelou, 2015).
- Disaggregated access metrics. Increasingly, access indicators are applied across socioeconomic and spatial categories—such as rural versus urban households, gendered access patterns, or the situation of refugees and informal settlers. Disaggregation provides a more nuanced understanding of inequality, as national statistics may obscure significant disparities (Sovacool et al., 2019).

A critical distinction also emerges when applying access metrics in the Global South versus the Global North. In the Global South, access indicators often focus on basic connections to electricity and clean cooking, as large population segments remain unserved or underserved. Reliability, affordability, and safety dimensions are increasingly integrated but are often secondary to the fundamental question of connection. In contrast, in the Global North, where near-universal electrification has long been achieved, access debates emphasize service quality, digital inclusion, and resilience. Here, metrics such as outage frequency, grid stability, and access to advanced energy services (e.g., broadband-enabled smart meters and electric mobility infrastructure) are more relevant for assessing equity. This divergence underscores the importance of context-sensitive measurement: while universal access remains a priority in many parts of the Global South, ensuring equitable participation in decarbonized, digitalized energy systems is a pressing challenge in the Global North (International Energy Agency, 2024; IEA et al., 2025).

Taken together, access metrics provide a multidimensional view of energy equity by considering not only whether households are connected but also the quality, reliability, and inclusiveness of the services they receive. This makes them central to evaluating both progress and persistent inequities in achieving universal modern energy access.

7.1.1.3 Distributional Burden Metrics

Distributional burden metrics are applied to assess how the financial costs and benefits of energy systems and policies are distributed across households and social groups. They are central to equity evaluations because they highlight whether energy transitions create regressive effects, where vulnerable households carry disproportionate burdens, or whether benefits are concentrated among wealthier groups. Unlike affordability metrics, which capture the financial strain at the household level, distributional metrics emphasize relative fairness in the allocation of costs and benefits within and across societies. The most popular distributional burden metrics can be singled out, such as:

- Household energy expenditure shares by income group. This metric compares the share of disposable income devoted to energy across different groups of households. It highlights whether low-income groups spend a higher proportion of their income on essential energy services, even if their absolute energy use is lower. Such comparisons reveal regressive patterns in energy pricing and policy outcomes.
- Subsidies allocation. Subsidy allocation measures how fuel or electricity subsidies are distributed across various population groups. While designed to support the poor, universal subsidies often benefit higher-income households more, as they consume larger amounts of energy. Analysis of subsidy incidence is therefore used to inform reforms toward more targeted social tariffs or direct transfers.
- Distributional effects of energy and carbon taxation. This metric evaluates how different policy instruments—such as carbon taxes, fuel excise duties, renewable levies on electricity bills, and clean energy incentive programs—affect household expenditures across socioeconomic groups. Indicators in this category are typically derived from household expenditure surveys and models that simulate the pass-through of policy costs and benefits into energy and transport prices.

- Distribution across gender and marginalized groups. This metric expands distributional analysis beyond income categories by focusing on inequalities linked to gender, age, and social identity. Examples include gender-disaggregated energy expenditure shares, which highlight differences between male- and female-headed households; urban–rural disparities in energy burdens, where rural households often face higher relative costs due to limited infrastructure; and vulnerability indices for marginalized groups, which integrate expenditure data with social categories such as disability, ethnicity, or migration status. Together, these indicators capture inequities that are overlooked when distributional analysis is restricted to income groups alone.

A distinction emerges in the application of distributional burden metrics between the Global South and the Global North. In the Global South, assessments focus primarily on subsidy incidence and reform, as universal subsidies remain a major policy instrument but often provide regressive outcomes. In the Global North, the emphasis lies on the distributional impacts of decarbonization policies, particularly carbon pricing, renewable support schemes, and clean energy incentives. Despite these contextual differences, distributional burden metrics in both regions provide crucial insights into who pays and who benefits from energy transitions, enabling the design of policies that combine environmental effectiveness with social fairness.

7.1.1.4 Procedural Equity Metrics

Procedural equity metrics capture the fairness of decision-making processes in energy systems, emphasizing whether different groups have equal opportunities to participate, influence, and be represented. Unlike affordability or distributional metrics, which mostly focus on financial outcomes, procedural metrics evaluate the inclusiveness, transparency, and accountability of governance. These measures are crucial for ensuring that energy transitions do not reproduce existing power imbalances or exclude marginalized voices. The most frequently applied procedural equity metrics include:

- Stakeholder participation rates. This indicator measures the extent to which diverse stakeholders—such as community organizations, NGOs, or consumer associations—are involved in decision-making processes. It is often quantified by the number and diversity of actors participating in consultations, public hearings, or policy forums.
- Representation of marginalized groups. This metric assesses whether vulnerable populations are formally represented in governance structures, regulatory boards, or participatory councils. Both the absolute number of representatives and their influence in shaping outcomes are considered.
- Transparency and accessibility of decision-making. Indicators in this category track the access of relevant information and whether consultation

outcomes are publicly reported. These measures highlight whether participation is substantive or merely symbolic.
- Quality of consultation processes. Beyond participation counts, this metric evaluates whether consultations are meaningful. Examples include the responsiveness of policymakers to stakeholder inputs and the presence of feedback mechanisms to incorporate concerns into final decisions.
- Legal and institutional guarantees of participation. This metric examines whether regulatory frameworks mandate public involvement—for example, through energy ombudsman services, guaranteed consultation procedures, or participatory budgeting for energy infrastructure. Such guarantees institutionalize procedural equity and reduce the risks of exclusion.

In the Global South, procedural metrics often focus on community involvement in electrification and renewable energy projects, highlighting whether local populations have genuine influence or are sidelined by top-down planning. In the Global North, attention is placed on formal consultation mechanisms, transparency, and stakeholder representation in regulatory and policy processes, with growing interest in ensuring the voices of vulnerable consumers and minorities are not marginalized. In both contexts, procedural equity metrics provide insight into the legitimacy of energy governance and the fairness of transition pathways.

7.1.1.5 Recognition Equity Metrics

Recognition equity metrics assess whether the diversity of social groups and their specific vulnerabilities are acknowledged in energy systems. While affordability, access, and procedural metrics focus on costs, services, and governance, recognition metrics emphasize visibility, identity, and respect—ensuring that the needs of marginalized populations are not overlooked or misrepresented. These metrics highlight whether disadvantaged groups are made visible in data, policy frameworks, and program design, providing a foundation for fairer distributional and procedural outcomes. The most frequently applied recognition equity metrics include:

- Disaggregated data availability. This metric evaluates whether energy-related statistics are broken down by gender, age, disability, ethnicity, or migration status. The absence of such disaggregation indicates invisibility and a lack of recognition in official monitoring systems.
- Policy acknowledgment of vulnerable groups. Recognition is reflected in whether energy strategies, regulations, or transition plans explicitly address the circumstances of groups such as women, elderly people, rural households, or indigenous communities. The presence of targeted provisions signals recognition at the policy level.
- Perception-based indicators. Surveys and participatory assessments capture whether individuals feel their needs, identities, and lived experiences

are respected in energy decision-making (Eurostat, 2025). These subjective metrics, while less standardized, add critical insights into recognition that cannot be derived from quantitative data alone.

A contextual difference emerges between the Global South and the Global North. In the Global South, recognition equity metrics often emphasize acknowledging the basic energy needs of groups such as women engaged in fuel collection, off-grid households, or indigenous communities affected by infrastructure projects. In the Global North, metrics focus more on the inclusion of categories such as disability, migration background, or aging populations in national energy strategies and welfare programs.

To consolidate the discussion of equity metrics, a comparative overview is provided in Table 7.1. While each metric group has distinct objectives and methodologies, their combined application highlights the multidimensional nature of equity in energy systems. Table 7.1 summarizes their core focus, the most frequently applied indicators, and contextual differences between the Global South and Global North.

Table 7.1 Comparative overview of equity metrics in energy systems. Source: developed by the author.

Equity dimension	*Core focus*	*Most frequently applied metrics*	*Global South / Global North differences*
Affordability	Financial burden of energy costs on households	• Household energy spending; • Energy expenditure as a share of income (energy burden); • Energy price levels and trends; • Price-to-income ratios • Volatility-adjusted energy prices; • Arrears on utility bills and disconnections	**South:** focus on affordability of minimum energy baskets and coping with price volatility (e.g., kerosene, LPG). **North:** focus on relative burdens, regressivity of tariffs, arrears, and disconnections.
Access	Availability, reliability, and quality of modern energy services	• Electrification rates; • Access to clean cooking fuels/technologies; • Service quality and reliability (outage frequency, hours of supply); • Multi-tier framework indicators (availability, affordability, safety); • Disaggregated access data (urban/rural, gender, etc.)	**South:** focus on basic access to electricity and clean cooking. **North:** emphasis on service quality, reliability, digital inclusion, and advanced services.

(continued)

Table 7.1 (continued)

Equity dimension	*Core focus*	*Most frequently applied metrics*	*Global South / Global North differences*
Distributional	Fair distribution of costs and benefits of energy systems and policies	• Distribution of carbon pricing impacts; • Allocation of renewable support costs; • Distribution of clean energy program benefits; • Gender-disaggregated energy expenditure shares; • Urban–rural disparities in energy burdens; • Vulnerability indices for marginalized groups	**South:** analysis of subsidies, fuel switching, and access to benefits for low-income or rural households. **North:** focus on regressivity of carbon pricing, renewable levies, and unequal uptake of efficiency/EV programs.
Procedural	Fairness of decision-making processes and governance structures	• Stakeholder participation rates; • Representation of marginalized groups; • Transparency and accessibility of decision-making; • Quality of consultation processes; • Legal and institutional guarantees of participation	**South:** community involvement in electrification and renewable projects, avoiding top-down exclusion. **North:** formal consultation mechanisms, transparency, and representation of vulnerable consumers.
Recognition	Visibility and acknowledgment of marginalized groups and their needs	• Disaggregated data availability (gender, age, disability, etc.); • Policy acknowledgment of vulnerable groups; • Perception-based indicators of recognition	**South:** emphasis on recognition of women's unpaid labor, Indigenous rights, and off-grid households. **North:** attention to disability, aging, migration, and minority groups in policies and welfare programs.

Taken together, the metrics illustrate how equity assessments extend beyond economic affordability to encompass access, distributional outcomes, governance processes, and the visibility of marginalized groups. This integrated perspective establishes the foundation for multidimensional assessment methodologies, which will be addressed in the following subsection.

7.1.2 Methodologies for Multidimensional Assessment

Single indicators provide useful insights into affordability, access, or distributional fairness, but they cannot capture the full complexity of equity in energy systems. Multidimensional methodologies have therefore been developed to assess equity across distributional, procedural, and recognition dimensions, linking outcomes to stakeholder participation and social vulnerability. Such approaches aim to provide a holistic picture of fairness in energy systems and to guide decision-making that balances competing social objectives.

7.1.2.1 Methodological Considerations

Although much of the literature refers to energy justice metrics, the methodological considerations are equally relevant for equity assessments, as both fields rely on similar tools to evaluate fairness. Baker et al. (2023), in their review of metrics for decision-making in energy justice, identify several key insights that directly inform equity methodologies.

First, metrics must be decision-relevant: their design should reflect the needs of policymakers, regulators, utilities, and communities and provide evidence that can directly guide planning, regulation, or investment.

Second, multidimensional methodologies must recognize the interdependencies between equity dimensions. Distributional outcomes are often shaped by procedural fairness, while recognition of marginalized groups influences both participation and benefit-sharing. Treating these domains in isolation risks overlooking critical interactions.

Third, the creation of composite or aggregated metrics raises normative trade-offs. Assigning weights to different equity dimensions—for instance, prioritizing distributional over procedural outcomes—inevitably involves value judgments. Transparency in these methodological choices is crucial to ensure both legitimacy and comparability.

Fourth, these methodologies face data, scale, and comparability challenges. High-resolution socio-demographic, spatial, and temporal data are often required to operationalize metrics, yet such data remain uneven or inaccessible across jurisdictions. This limits the potential for standardized assessments and cross-country benchmarking.

Finally, fitness for scale is essential: multidimensional methodologies must be adaptable across levels of analysis, from local communities and distribution grids to national systems, and across temporal horizons ranging from hourly dynamics to long-term transitions (Baker et al., 2023).

While Baker et al. (2023) emphasize methodological principles for how equity assessments should be designed and implemented, Ren et al. (2025) provide systematic evidence on the methods most frequently applied in practice, offering a complementary perspective on operational tools. The most frequently applied methodological approaches for multidimensional energy equity assessment include the following (Ren et al., 2025):

- Gini coefficient. The Gini is the most widely applied inequality metric, commonly used to examine disparities in household energy expenditure, electricity consumption, or access to modern energy services. Its main advantage lies in simplicity and communicability, although it cannot be decomposed into subgroup inequalities and may mask underlying structural disparities (Jacobson et al., 2005; Yitzhaki and Schechtman, 2013). For these reasons, the Gini coefficient is often used as a first descriptive indicator in energy equity studies but is most effective when complemented by measures such as the Atkinson index or Generalized Entropy indices that allow deeper distributional analysis.

- Generalized Entropy indices are increasingly applied in energy equity assessments due to their flexibility and decomposability. By adjusting a sensitivity parameter, these measures can emphasize different parts of the distribution. For example, Theil's L (mean log deviation) is more sensitive to disparities at the lower end, making it useful for highlighting inequities affecting the poorest households, while Theil's T places greater weight on disparities at the upper end, reflecting concentration among wealthier groups (Theil, 1967; 1979). This parameterization allows analysts to tailor the focus of inequality measurement depending on policy priorities. Generalized Entropy indices also have the advantage of being fully decomposable, allowing inequality to be separated into within-group and between-group components—for example, comparing urban and rural households or income quintiles—thereby revealing structural patterns of inequity in energy systems. Despite these advantages, Generalized Entropy indices are not without limitations. Their results can vary substantially depending on the parameter chosen, which may lead to divergent conclusions across studies. Furthermore, because they emphasize relative disparities, they may obscure the absolute forms of hardship faced by households with limited or no access to essential energy services (Ren et al., 2025). For this reason, they are most effective when combined with other indicators that capture absolute deprivation, such as access metrics or poverty thresholds. Overall, Generalized Entropy indices provide a versatile and decomposable framework for assessing equity in energy systems, but their interpretation requires careful attention to parameter choice and contextual conditions.
- Atkinson index. The Atkinson index is a distributional measure frequently applied in energy equity assessments. Its main advantage is the ability to weight inequalities at different points of the distribution, making it particularly suitable for identifying whether low-income or otherwise vulnerable households carry disproportionate energy burdens. In practice, it has been used to evaluate the distributional impacts of electricity tariffs, fuel price changes, and subsidy reforms, thereby revealing trade-offs between efficiency and fairness in energy systems. The index is especially valuable in contexts where equity assessments explicitly prioritize the worst-off, as the sensitivity parameter allows greater weight to be assigned to disparities at the lower end of the distribution. At the same time, outcomes are sensitive to the parameter chosen, and transparency in its selection is essential to ensure comparability and legitimacy of results.
- Variance-based measures. Variance-based measures, including standard deviation and coefficient of variation, represent some of the earliest

approaches to inequality analysis and are still applied in energy equity assessments. Their main advantage lies in computational simplicity and the ability to provide a quick overview of the dispersion of household energy expenditures, electricity consumption, or fuel use around the mean (Ren et al., 2025). The coefficient of variation is particularly useful because it normalizes inequality relative to the mean, facilitating comparisons across countries or time periods with different income and consumption levels. However, variance-based measures are highly sensitive to extreme values, which may distort results in the presence of outliers, and they do not provide information about the location of inequality within the distribution. Moreover, unlike Generalized Entropy indices, they cannot be decomposed into within-group and between-group components, limiting their usefulness for subgroup analysis such as urban–rural or income-based comparisons. For these reasons, variance-based measures are generally used as complementary indicators, providing a simple picture of dispersion but requiring support from more sophisticated indices to capture structural inequities.

- Fairness indices from other fields. In addition to traditional inequality measures, several fairness indices originally developed in economics, political science, and computer science are beginning to be applied in energy equity assessments. These indices are designed to evaluate whether burdens and benefits are shared in ways that align with principles of fairness, even if they were not initially created for energy contexts (Ren et al., 2025). Examples can be indices for Quality of Service (Hoßfeld et al., 2016) and Quality of Experience (Moret and Pinson, 2019). The main advantage of such indices is their ability to explicitly integrate normative fairness criteria into technical and policy assessments, providing novel perspectives on distributive and procedural justice in energy systems. However, their application to energy equity remains at an early stage, and methodological challenges persist. These include the need to adapt definitions of fairness to energy-specific contexts, to reconcile different fairness principles that may produce conflicting outcomes, and to ensure that quantitative results remain meaningful for policy and governance. Despite these limitations, borrowing fairness indices from other disciplines broadens the methodological toolbox for energy equity analysis and supports more innovative approaches to evaluating just and inclusive transitions.

However, general inequality metrics offer insights into distributional fairness, but they do not reflect procedural or recognition-based equity dimensions, which are essential for a comprehensive understanding. This limitation persists even when such metrics are combined.

This list of methodological tools can be further supplemented by several broader practices that are frequently applied in the literature when addressing questions of fairness in the energy sector.

- Energy poverty or vulnerability indices. The prevailing approach is to rely on energy poverty or vulnerability indices, which, although designed to capture deprivation, are often used as proxies for equity analysis. Examples include the Multidimensional Energy Poverty Index (MEPI) (Nussbaumer et al., 2012), national-level indices, and simplified affordability thresholds such as the 10% rule, the After Fuel Cost (AFC) indicator, or the Low-Income High-Cost (LIHC) measure (Hills, 2012). While these tools have become mainstream in research and policy, they are essentially focused on poverty and vulnerability, not on equity in its fuller sense.
- Case-specific indices. Despite the fact that some case-specific indices do exist—such as the Energy Equity Index within the World Energy Council's "Energy Trilemma Index"—equity in this case is defined primarily in terms of affordability and accessibility of energy supply (World Energy Council, 2023). This approach, however, is more closely aligned with the measurement of energy access and poverty than with a comprehensive assessment of social equity.
- New composite frameworks. To address these gaps, some researchers have turned to energy justice metrics, which explicitly seek to capture fairness across distributional, procedural, and recognition dimensions (Jenkins et al., 2016). Yet, methodological fragmentation remains: most studies construct their own composite frameworks or "justice indices," combining indicators of affordability, access, participation, and recognition in varied ways, making comparability difficult.
- Context-specific frameworks. Finally, the literature also offers context-specific innovations, including equity-sensitive renewable energy indices and gender equity metrics assessing participation in clean energy employment or access to modern energy services. These contributions highlight important equity aspects but remain scattered across disciplines and lack standardization.

Taken together, it can be stated that while inequality metrics and energy poverty indices remain dominant in practice, and justice frameworks offer conceptual breadth, there is no widely accepted, multidimensional methodology dedicated to social equity in energy systems. Addressing this gap requires not only the development of standalone equity indices but also the integration of equity-sensitive indicators into established sustainability, energy security, and transition monitoring frameworks to ensure that just transition principles are embedded into policy and governance.

In addition to specific indices, methodological frameworks also classify metrics according to their role in the policy cycle. For example, Tarekegne et al. (2021) distinguish between (i) target population identification metrics (e.g., burden thresholds, vulnerability indices), (ii) investment decision-support metrics (distribution of costs and benefits of tariffs or infrastructure), and (iii) program impact metrics (evaluating whether interventions achieve equitable outcomes). This typology clarifies how methodological choices align with policy objectives.

The methodological landscape for assessing social equity in the energy sector remains diverse, dynamic, yet fragmented. Existing approaches—ranging from single inequality indicators and decomposable statistical indices to multidimensional and context-specific frameworks—each contribute valuable but partial perspectives. While classical inequality metrics such as the Gini, Theil, and Atkinson indices offer rigorous insights into distributional fairness, they fail to capture procedural and recognition dimensions that are central to a holistic understanding of equity. Similarly, energy poverty and vulnerability indices remain useful for identifying deprivation but are insufficient for assessing fairness in participation, representation, or benefit-sharing. Emerging fairness indices and energy justice frameworks attempt to bridge these gaps, but they currently lack conceptual consistency and methodological standardization across contexts.

This diversity of approaches highlights both the progress and the limitations of current methodological practice. On the one hand, the growing use of decomposable indices, multi-criteria frameworks, and participatory approaches reflects a maturing recognition that equity is multidimensional and context-dependent. On the other hand, the coexistence of overlapping tools without harmonized definitions or weighting principles undermines comparability and limits policy uptake. Moreover, persistent data asymmetries—particularly the absence of high-resolution socio-demographic, spatial, and temporal data—constrain the operationalization of equity metrics across scales.

Looking ahead, the field requires a systematic integration of equity-sensitive metrics into established policy assessment instruments, including sustainability evaluations, energy security indices, and transition monitoring frameworks. Future methodological work should focus on developing transparent, multidimensional, and scalable equity indices that link distributional outcomes with procedural inclusiveness and recognition of social diversity. Such progress would not only enhance the analytical robustness of energy equity assessments but also strengthen their policy relevance, ensuring that equity principles are embedded throughout the design, implementation, and evaluation of energy systems.

7.1.3 Data Gaps and Comparability Challenges

Advancing equity assessment in energy systems requires reliable, disaggregated, and comparable data. Yet the development and application of equity metrics remain constrained by significant data gaps and methodological inconsistencies, which undermine both national policy evaluation and international benchmarking. These limitations affect not only affordability and access indicators, but also more complex distributional, procedural, and recognition metrics. The main challenges include:

- Scarcity of high-resolution data. Many equity metrics require information on household-level energy use, socio-demographic characteristics, and spatial patterns of infrastructure. Such datasets are often incomplete, unavailable, or restricted due to privacy regulations. For example, expenditure surveys frequently lack detail on fuel types or household energy services, making it difficult to assess differentiated burdens across groups (Heindl and Schüßler, 2015). Procedural and recognition equity are particularly underrepresented, since data on participation in decision-making or the situation of marginalized groups is rarely collected in standardized surveys (Baker et al., 2023).
- Lack of comparability across jurisdictions. Even where data exist, differences in definitions, coverage, and collection methods prevent meaningful cross-country or cross-regional comparisons. Energy burden metrics vary in how household income and expenditures are defined, while access indicators may apply different thresholds for "modern energy services." Without harmonized methodologies, benchmarking equity outcomes across contexts remains problematic (Barlow et al., 2022).

Table 7.2 Key data gaps and comparability challenges in energy equity assessments. Source: developed by the author based on Heindl and Schüßler (2015), Barlow et al. (2022), and Baker et al. (2023).

Challenge	*Description*	*Implications for equity metrics*	*Potential responses*
Scarcity of high-resolution data	Household-level energy use, socio-demographic, and spatial data are often incomplete, unavailable, or restricted due to privacy.	Limits ability to disaggregate burdens across groups and assess procedural/recognition equity.	Invest in improved household surveys; integrate administrative and utility datasets.
Lack of comparability	Definitions of income, expenditure, or "modern energy access" vary across countries and surveys.	Prevents benchmarking and international policy learning.	Develop standardized guidelines and harmonized definitions for equity indicators.
Scale mismatches	National datasets obscure local disparities; utility-level data are often proprietary.	Underrepresents inequalities at community or regional levels.	Encourage open-access utility data; promote multi-scalar data collection.
Bias and exclusion	Vulnerable groups (migrants, informal settlers, off-grid households) are underrepresented in official data.	Creates blind spots, undermining policy targeting and recognition equity.	Ensure inclusion of marginalized groups in surveys; design participatory data approaches.

- Scale mismatches. Data collected at the national level may obscure local disparities, while utility-level datasets are often proprietary and inaccessible to researchers. This limits the ability to analyze equity at scales where disparities are most pronounced, such as between urban and rural communities or across different socioeconomic groups.
- Bias and exclusion in datasets. Vulnerable groups such as informal settlers, migrants, or off-grid households are frequently underrepresented in official statistics, leading to systematic blind spots in equity assessments (Baker et al., 2023). This not only weakens the validity of metrics but also risks reproducing inequities by failing to recognize those most affected.

These challenges highlight the need for investments in data infrastructure and standardization. Improved household surveys, integration of administrative and utility records, and the development of open-access databases could help overcome current gaps. Equally important is the creation of standardized guidelines for equity metrics, which would allow comparisons across jurisdictions while retaining sensitivity to local contexts. Table 7.2 provides a structured overview of the main challenges, their implications, and possible responses.

7.2 Policy Instruments and Governance Mechanisms for Advancing Equity

7.2.1 Regulatory and Institutional Frameworks

The advancement of social equity in the energy sector fundamentally depends on the quality, coherence, and inclusiveness of its regulatory and institutional environment. Regulatory and institutional frameworks establish the legal and procedural foundations that determine how energy systems operate—defining who participates in decision-making, how responsibilities and benefits are distributed, and whose needs and values are recognized in policy outcomes. In this sense, they form the structural core of equitable energy governance, providing the conditions under which distributive, procedural, recognition, and other justice dimensions can be effectively realized in practice.

Regulatory frameworks specify the legal obligations and rights that govern energy markets, encompassing tariff design, universal service obligations, environmental and safety standards, and consumer protection. Institutional frameworks, in turn, embody these principles within administrative and governance structures such as ministries, regulatory authorities, coordination councils, and participatory platforms. Together, they ensure that equity objectives are not treated as aspirational or secondary but as integral, enforceable, and measurable components of energy governance.

Historically, energy regulation has primarily focused on economic efficiency, reliability, and competition. However, the acceleration of the low-carbon transition, the decentralization of energy systems, and the increasing social salience of energy poverty and vulnerability have broadened the regulatory agenda.

Contemporary governance paradigms increasingly emphasize inclusiveness, transparency, and accountability, recognizing that energy transitions must be not only technically and economically feasible but also socially legitimate. This shift has given rise to a multilayered governance architecture in which social equity is addressed through interconnected legal, institutional, and participatory mechanisms.

For example, in the European context, this transformation is particularly evident. The European Union (EU) has progressively integrated social considerations into energy and climate legislation, incorporating explicit provisions on energy poverty, protection of vulnerable consumers, and public participation. The EU Member States have subsequently aligned their national frameworks with EU documents, while international organizations such as the United Nations, the International Energy Agency (IEA), and the Organisation for Economic Co-operation and Development (OECD) have developed complementary global frameworks promoting fairness, inclusiveness, and just transitions.

To clarify analytical structure, regulatory and institutional frameworks can be grouped into five interrelated categories, each contributing in distinct ways to equity (Table 7.3). To illustrate how social equity is operationalized through governance structures, Table 7.3 presents a typology of regulatory and institutional frameworks commonly applied within the EU and its Member States. The EU provides a particularly instructive context, as its energy governance model combines supranational legislation, national implementation, and regional participation—offering a comprehensive example of how equity principles are embedded across multiple levels of policy design and execution. While the typology draws primarily on the EU framework, its categories also capture broader international tendencies relevant to equity-oriented energy governance.

Regulatory and institutional frameworks interact and mutually reinforce one another, forming a layered governance architecture that links normative principles with binding legal obligations, administrative structures, cross-sectoral coordination, and participatory mechanisms. Together, they demonstrate that equitable energy governance emerges not from isolated interventions but from the coherence of interdependent frameworks that align strategic objectives with implementation practices.

Normative and strategic frameworks establish social equity as a fundamental objective of energy policy. Although generally not legally binding, they provide strategic direction, policy legitimacy, and a benchmark for institutional accountability. At the global level, United Nations SDG 7 positions universal access to affordable, reliable, sustainable, and modern energy as a central pillar of sustainable development (United Nations, 2015a), while the Paris Agreement (United Nations, 2015b) explicitly calls for a "just transition" that safeguards workers and communities during decarbonization. Within the EU, the European Pillar of Social Rights (European Parliament et al., 2017) reinforces the social dimension of sustainability by affirming principles of equality,

Table 7.3 Regulatory and institutional frameworks for social equity in the energy sector. Source: developed by the author.

Category	*Primary function*	*Examples (EU context)*	*Main contribution*
Normative and strategic frameworks	Establish guiding values and policy legitimacy for fairness, inclusiveness, and justice in energy governance.	• UN's SDG 7; • Paris Agreement • European Pillar of Social Rights; • European Green Deal; • European National Energy and Climate Plans	Strengthens recognition and procedural equity by legitimizing inclusiveness, recognition, and participation as policy principles.
Regulatory frameworks	Translate equity and justice principles into binding legal and market rules.	• EU Clean Energy for All Europeans Package; • Electricity Directive (EU) 2019/944; • Energy Efficiency Directive; • National energy laws and acts	Promotes distributional equity by ensuring affordable, reliable, and nondiscriminatory access to energy services.
Institutional governance frameworks	Define administrative and organizational structures for implementation, monitoring, and enforcement.	• National Regulatory Authorities; • EU Agency for the Cooperation of Energy Regulators (ACER); • Just Transition Task Forces and interministerial committees	Supports procedural and recognition equity through transparent, accountable, and inclusive governance arrangements.
Coordination and integration frameworks	Align energy, social, welfare, housing, and regional policies to achieve coherent equity outcomes.	• EU Just Transition Mechanism; • European Social Climate Fund; • EU Cohesion Policy	Integrates distributional, procedural, and recognition equity by ensuring multi-sectoral coherence and balanced regional development.
Adaptive and participatory frameworks	Foster flexibility, stakeholder engagement, local empowerment, and community-energy sovereignty.	• Adaptive regulatory mechanisms, such as regulatory sandboxes; • Institutionalized participatory planning councils • Community energy initiatives and energy cooperatives	Enhances procedural and recognition equity by embedding stakeholder participation, promoting inclusiveness, and strengthening self-determination, social learning, and adaptive capacity in energy governance.

inclusion, and fair working conditions. The European Green Deal (European Commission, 2019) translates these values into a comprehensive growth strategy, embedding social cohesion and fairness within the decarbonization agenda. Correspondingly, National Energy and Climate Plans (European Commission, 2025a) operationalize these principles at the Member States level, linking energy affordability, inclusiveness, and social justice to national policy objectives.

Regulatory frameworks convert normative principles into enforceable rules. For example, the EU Clean Energy for All Europeans Package is an example, comprising several directives and regulations that embed consumer rights, protection for vulnerable users, and universal service obligations into energy markets. Under the Electricity Directive (EU) 2019/944 (European Parliament and Council, 2019), the EU countries must protect vulnerable consumers and guarantee nondiscriminatory network access. Meanwhile, national legislation—such as social-tariff regimes or subsidized electricity pricing schemes—translates those mandates into domestic entitlement. As these frameworks primarily promote distributional equity, ensuring affordable energy access for vulnerable populations, they also include procedural safeguards (e.g., complaint mechanisms, oversight), which contribute to participatory fairness.

Institutional governance frameworks operationalize and monitor the implementation of regulatory mandates, ensuring that equity objectives are embedded in the administrative and oversight processes of the energy sector. National Regulatory Authorities (NRAs) such as VERT in Lithuania, Ofgem in the UK, or CRE in France set tariff rules, monitor compliance, and adjudicate disputes. At the European level, the Agency for the Cooperation of Energy Regulators (ACER) coordinates these agencies, especially in cross-border energy markets. Many jurisdictions also establish Just Transition Task Forces or interministerial councils to manage the socioeconomic dimensions of energy decarbonization, ensuring that local stakeholders, labor, and community interests are represented. These institutional arrangements serve procedural justice and recognition justice.

Coordination and integration frameworks link energy governance with interconnected policy areas such as housing, welfare, regional development, and climate action. Their primary function is to ensure that these domains work in harmony rather than in isolation, aligning social, environmental, and economic objectives to avoid conflicting measures and to strengthen synergies that support an equitable and inclusive energy transition. The EU Just Transition Mechanism (2021–2027) exemplifies this approach by linking decarbonization funding with social and regional fairness, particularly in carbon-intensive regions (European Parliament and Council, 2021a), while the European Social Climate Fund integrates climate and social policies by supporting vulnerable households and communities affected by carbon-pricing measures (European Parliament and Council, 2023). EU Cohesion Policy further promotes balanced regional development to mitigate spatial disparities arising from the energy transition (European Commission, 2021). At the national level, cross-ministerial committees and energy-welfare coordination bodies integrate social protection measures with energy and climate regulation, ensuring that affordability, inclusion, and sustainability objectives are pursued in parallel. By bridging sectoral divides and promoting inter-institutional collaboration, coordination, and integration frameworks operationalize distributional, procedural, and recognition equity through coherent, multi-level policy design and implementation.

Adaptive and participatory frameworks strengthen flexibility, inclusiveness, and stakeholder engagement in the governance of modern energy systems, which are becoming increasingly decentralized, digitalized, and complex. Their primary role is to ensure that policy and regulatory environments remain responsive to social needs and technological change, while empowering citizens and communities to take an active role in energy production and decision-making. National and regional initiatives that support community-energy projects and energy cooperatives illustrate this approach, enabling collective ownership, local benefit-sharing, and self-governance of energy resources. Within the EU, these efforts are reinforced through the implementation of the Renewable Energy Directive (EU) 2018/2001 (European Parliament and Council, 2018a) and the Electricity Directive (EU) 2019/944 (European Parliament and Council, 2019), which formally recognize Renewable Energy Communities and Citizen Energy Communities as new legal entities that institutionalize public participation and local empowerment in the energy sector. Additionally, some countries have introduced regulatory sandboxes—temporary frameworks that allow controlled experimentation with new energy business models, digital solutions, and market arrangements. While primarily designed to foster innovation, they also contribute to adaptive governance by enabling regulators to learn from emerging practices and adjust rules in response to technological and societal change (Schittekatte et al., 2021; European Commission, 2023). Together, these frameworks advance procedural and recognition equity by embedding participation, transparency, and inclusiveness into energy governance, fostering greater community agency, adaptive governance capacity, and long-term social legitimacy in the energy transition.

This governance architecture illustrates how regulatory and institutional arrangements operate together to embed social equity across all levels of the energy system. The normative and strategic layer establishes legitimacy and policy direction; regulatory mechanisms translate equity principles into binding obligations; institutions oversee implementation and accountability; coordination mechanisms align cross-sectoral and territorial objectives; and adaptive and participatory arrangements link governance processes to citizens and communities. These interdependent layers demonstrate that equitable energy transitions can only be achieved when governance arrangements are coherent, well-coordinated, and implemented in a logical sequence from policy design to execution.

7.2.2 Economic and Financial Instruments for Equity

Economic and financial instruments are the principal tools through which the principles of energy justice and social equity are implemented across the entire energy sector. They determine how the financial responsibilities and benefits arising from energy production, distribution, and consumption are shared among social groups, regions, and generations. While regulatory and institutional frameworks establish the legal and procedural foundations for fairness, economic and financial tools translate these commitments into practice through pricing policies, fiscal transfers, taxation schemes, subsidies, and targeted investment programs. Their design influences affordability, accessibility, and reliability of energy services, as well as the extent to which vulnerable consumers, disadvantaged regions, and strategic sectors are supported within the overall energy system.

Beyond redistribution, these instruments also perform stabilizing and corrective functions: they mitigate income shocks caused by energy price volatility, compensate for environmental and social externalities, and guide investment toward socially and environmentally desirable outcomes. When designed in accordance with justice and equity principles, economic and financial mechanisms enhance public trust, maintain policy legitimacy, and ensure that energy systems operate not only efficiently and securely but also fairly.

To provide a clearer analytical overview, economic and financial instruments can be grouped into several interrelated categories according to their primary function and type of operation. It allows illustrating the different mechanisms through which equity objectives are pursued in the energy sector—ranging from redistributive support and market regulation to investment promotion and international cooperation. Table 7.4 presents a typology of these instruments, highlighting how diverse policy tools can advance fairness and inclusiveness across multiple scales. The classification draws on both global and EU experience, as the EU provides one of the most comprehensive examples of how economic and financial measures can be systematically aligned with social equity goals.

As it can be seen in Table 7.4, economic and financial instruments encompass a broad spectrum of mechanisms operating at national, regional, and international levels. Their equity implications depend on design parameters, implementation capacity, and interaction with wider governance frameworks.

7.2.2.1 Redistributive and Social Protection Instruments

Redistributive and social protection instruments constitute the most immediate policy tools for safeguarding affordability and ensuring universal access to essential energy services. Their central function is to redistribute the financial burden of energy costs across social groups, protecting low-income and vulnerable households from price volatility and excessive expenditure. These mechanisms operate primarily at the national or subnational level, situated at the intersection of energy and social policy, and aim to guarantee that energy remains a basic, non-exclusionary service.

Within the EU, redistributive measures are embedded in long-standing national frameworks that integrate social energy tariffs, energy-voucher schemes, and targeted income-support transfers. These are frequently complemented by Public Service Obligation funds, which ensure the continuity of supply and uphold universal service obligations (Karpavicius and Balezentis, 2022). The forthcoming Social Climate Fund (2026–2032) represents an additional layer of protection at the EU level, designed to mitigate the

Table 7.4 Economic and financial instruments for social equity in the energy sector. Source: developed by the author.

Category	*Primary function*	*Illustrative examples (EU / global context)*	*Contribution to equity and justice*
Redistributive and social protection instruments	Reduce affordability gaps and protect vulnerable households from excessive energy costs.	EU: Social energy tariffs, Public Service Obligation funds, energy-voucher programs, targeted income-support schemes under the Social Climate Fund. Global: Targeted fuel subsidies, lifeline tariffs.	Advance distributive and recognition justice by ensuring affordable access for low-income or vulnerable consumers and mitigating short-term energy hardship.
Market-based instruments	Internalize social and environmental costs while creating incentives for efficient and fair energy consumption and production.	EU: EU Emissions-Trading System (EU ETS), national carbon taxes, feed-in tariffs and contracts-for-difference, energy-efficiency obligation schemes.Global: Clean Development Mechanism, national emissions-trading schemes	Promote distributional equity when revenues are recycled progressively; support restorative justice when proceeds finance compensation or social investment.
Public investment and funding mechanisms	Channel public capital toward inclusive infrastructure, modernization, and access expansion.	EU: Just Transition Mechanism, European Social Climate Fund, Cohesion Policy funds, European Investment Bank lending, national energy-efficiency and renovation fundsGlobal: World Bank Energy Sector Management Assistance Program, Asian Development Bank Energy Transition Mechanism, Green Climate Fund	Support recognition and procedural justice by financing access in underserved regions and enabling local participation in project design and monitoring.
Compensatory and corrective mechanisms	Address structural inequalities and legacy effects of past energy policies or industrial decline.	EU: Just Transition Mechanism, Modernization Fund, national compensatory schemes Global: Loss and Damage Fund, Climate Investment Funds, Just Energy Transition Partnerships, Global Environment Facility, environmental remediation and reskilling funds	Serve restorative justice by repairing historical or structural harm and providing social support during structural adjustments.

distributive impacts of carbon pricing by providing direct income support and financing energy-efficiency renovations for low-income households (European Parliament and Council, 2023). Together, these measures operationalize the principles of distributive and recognition justice by embedding affordability and inclusion within domestic energy-governance systems.

Comparable approaches exist beyond Europe. For example, lifeline tariffs in South Africa and India and targeted fuel-subsidy schemes in Mexico and Indonesia perform similar redistributive functions by maintaining affordability for basic energy consumption while gradually reforming inefficient universal subsidies. In several Latin American and African countries, social assistance programs—such as conditional cash transfers or income-support schemes—are coordinated with energy-affordability policies, enabling low-income households to automatically benefit from social electricity tariffs or other targeted subsidies. While these mechanisms vary in design, they share a common objective: to stabilize household energy access and prevent energy deprivation, especially during periods of market fluctuation or economic stress.

The effectiveness of redistributive instruments depends on precise targeting, stable funding, and strong coordination between social and energy policy domains. Well-designed schemes can alleviate energy poverty and reinforce social resilience without undermining market efficiency. Conversely, poorly targeted or inadequately selected measures may create fiscal imbalances, distort price signals, or weaken incentives for energy savings. Ultimately, redistributive and social protection instruments remain indispensable for addressing short-term affordability challenges, but their long-term contribution to equity and poverty alleviation requires integration with broader structural policies that enhance efficiency, income security, and sustainable energy access (Jové-Llopis et al., 2025).

7.2.2.2 Market-based Instruments

Where redistributive instruments address short-term affordability and protection from price volatility, market-based instruments operate through pricing and trading mechanisms that internalize environmental and social costs into energy markets. These instruments influence producer and consumer behavior by adjusting price signals and creating fiscal revenues that can be recycled toward social and environmental objectives. In contrast to redistributive tools, which directly transfer income or subsidize consumption, market-based instruments steer market behavior through price-based measures—such as carbon taxes and emissions-trading systems—or quantity-based mechanisms, including energy-efficiency obligations and tradable certificates. When designed with explicit equity safeguards, they can combine economic efficiency with social fairness, ensuring that the costs of energy production and decarbonization are shared proportionally and do not fall disproportionately on low-income or vulnerable groups.

Within the EU, the EU ETS serves as the cornerstone of market-based climate and energy policy. Although its primary aim is to reduce greenhouse-gas

(GHG) emissions cost-effectively, the auctioning of emission allowances provides significant revenues that Member States increasingly direct toward equity-enhancing measures (European Parliament and Council, 2003). Complementary national instruments, such as carbon taxes applied in non-ETS sectors, generate fiscal revenues that may be recycled through redistributive measures, including lump-sum rebates to households, reductions in social contributions, or targeted transfers for energy-poor consumers. While these measures are not market instruments in themselves, their design determines whether carbon-pricing frameworks achieve progressive or regressive social outcomes (European Commission, 2025b). These mechanisms demonstrate how fiscal policy can operationalize distributive and recognition justice within a market framework, balancing efficiency with social inclusion.

Additional market-based approaches include energy-efficiency obligation schemes, which require energy suppliers to deliver specific end-use savings. In several EU Member States—such as France, Italy, and the United Kingdom—these obligations include priority actions for energy-poor or vulnerable households, ensuring that efficiency gains benefit those most in need. Likewise, feed-in tariffs, feed-in premiums, and contracts for difference, traditionally introduced to support renewable generation, can be designed to include social or regional criteria, such as local-employment requirements or community participation, thereby reinforcing procedural and distributive equity in energy-investment processes.

Globally, comparable policy architectures have evolved in diverse governance contexts. For example, national emissions-trading systems in New Zealand, the Republic of Korea, and China illustrate how cap-and-trade mechanisms are being adapted beyond the EU, with their equity outcomes shaped by allocation rules and revenue-use strategies. Under the Kyoto Protocol, the Clean Development Mechanism provided an early model for transferring low-carbon investment toward developing countries (United Nations, 1998; Huang and Barker, 2012), while the Paris Agreement's Article 6 framework seeks to enhance fairness, transparency, and local benefits in international carbon markets (United Nations, 2015a). In many countries (e.g., Canada, Japan, South Africa), carbon-pricing revenues have been explicitly earmarked for energy-efficiency programs, social rebates, and rural-electrification funds, illustrating how market instruments can advance energy equity when revenue-use criteria are explicitly defined (Marten and van Dender, 2019; Onolemhemhena and Bello, 2024). However, empirical evidence indicates that instruments such as carbon pricing can contribute to emission reductions, although their effectiveness largely depends on the adequacy of price signals, the strength of regulatory enforcement, and the extent to which they are integrated with complementary policies in renewable energy deployment and energy-efficiency improvements (Okedele et al., 2024).

Therefore, the contribution of market-based instruments to social equity ultimately depends on their governance and revenue-allocation design. Key determinants include the share of proceeds channeled toward vulnerable consumers, the fairness of cost-recovery mechanisms, and the transparency of exemption or compensation schemes. When these parameters are guided by justice and equity principles, market-based policies can reinforce distributive, procedural, and recognition justice, translating efficiency into socially balanced outcomes. Conversely, poorly regulated or regressive designs risk amplifying inequality by shifting transition costs onto those least able to pay. Embedding clear equity objectives and verifiable targeting mechanisms into carbon-pricing, trading, and efficiency-obligation frameworks is therefore essential to ensure that market mechanisms contribute to both economic and social sustainability.

7.2.2.3 Public Investment and Funding Mechanisms

While market-based measures operate primarily through pricing mechanisms, public investment and funding mechanisms address long-term structural inequities by directing fiscal resources to areas where market forces alone cannot ensure social inclusion. These mechanisms constitute a significant policy pillar for advancing equity in the energy sector. Unlike market-based or redistributive tools, they operate through the direct allocation of public resources—from national budgets, dedicated funds, or international financial institutions—to support infrastructure, innovation, and social programs that deliver equitable access to energy. Their central function is to reduce systemic disparities by ensuring that communities, regions, and social groups facing structural disadvantages are not excluded from the benefits of modern energy systems.

Within the EU, public investment in equity-sensitive energy policy is channeled through an integrated financial framework that combines territorial, social, and climate objectives. The Just Transition Mechanism represents the EU's flagship initiative linking decarbonization with distributive and restorative justice. Through its three pillars—the Just Transition Fund, the InvestEU Just Transition Scheme, and the Public Sector Loan Facility—the Just Transition Mechanism supports the socioeconomic restructuring of regions and workers affected by the phase-out of carbon-intensive industries, financing reskilling, diversification, and clean-energy infrastructure (European Parliament and Council, 2021a; 2021b; 2021c). Although it also serves compensatory functions by supporting communities facing industrial decline, the Just Transition Mechanism is primarily a forward-looking investment framework aimed at ensuring territorial cohesion and inclusive regional growth.

Complementing this, the European Social Climate Fund (2026–2032) is designed to mitigate the social impacts of carbon pricing by providing direct income support and financing building renovation and sustainable mobility for vulnerable households (European Parliament and Council, 2023). Together, these instruments institutionalize equity objectives within the EU's energy-transition funding architecture, transforming climate finance into a mechanism for social protection and territorial development.

Broader regional and cohesion policies reinforce these efforts. The Cohesion Policy Funds remain the EU's principal instruments for reducing regional disparities, financing energy-efficiency retrofits, renewable-energy deployment, and sustainable infrastructure in less-developed regions (European Parliament and Council, 2021d). The European Investment Bank complements these funds by providing large-scale lending and guarantees for projects that combine climate objectives with social inclusion, such as district-heating modernization, social-housing renovation, and low-income household efficiency upgrades. At the national level, dedicated energy-efficiency and renovation funds translate EU and national objectives into direct support for households, helping to reduce energy poverty and improve living conditions in vulnerable communities.

Beyond Europe, global financial institutions play a central role in advancing equity through public investment. For example, the World Bank's Energy Sector Management Assistance Program supports developing countries in expanding affordable and sustainable energy access, with an explicit focus on gender inclusion and social vulnerability (World Bank, 2024). Similarly, the Asian Development Bank's Energy Transition Mechanism mobilizes public and private finance to facilitate equitable coal phaseout and clean-energy deployment, advancing restorative and distributive justice in Asia's emerging economies (Asian Development Bank, 2025). The Green Climate Fund, established under the Paris Agreement, further enhances global fairness by providing concessional finance for low-carbon, climate-resilient projects that prioritize vulnerable communities and countries with limited access to private investment (United Nations, 2012).

Collectively, various public investment and funding mechanisms demonstrate how public finance can translate justice principles into measurable outcomes. By embedding distributive, recognition, and restorative equity into investment criteria, public funding mechanisms ensure that energy transitions—whether national or global—serve as vehicles for social inclusion and sustainable development. Their long-term effectiveness, however, depends on transparent governance, equitable targeting, and participatory oversight.

7.2.2.4 *Compensatory and Corrective Mechanisms*

While redistributive, market-based, and investment instruments aim to prevent inequities ex-ante, compensatory and corrective mechanisms address those that persist or newly emerge during energy-system transformations. Their function is remedial: to offset social, regional, or sectoral losses associated with structural change, energy price reforms, or the decarbonization of carbon-intensive industries. At their core, compensatory and corrective instruments embody the principles of restorative and recognition justice—acknowledging past or ongoing inequities and restoring the social, economic, or environmental balance disturbed by energy transitions or market restructuring. In operational terms, they provide financial support, reskilling opportunities, and targeted reinvestment in affected regions and sectors. These measures are particularly relevant in contexts where the decarbonization of coal, oil, and gas industries causes regional economic contraction, unemployment, or reduced fiscal capacity.

Within the EU, this logic is institutionalized through the Just Transition Mechanism, which supports territories and workers adversely affected by the shift to a climate-neutral economy. With its pillars the Just Transition Fund, the InvestEU Just Transition Scheme, and the Public Sector Loan Facility (European Parliament and Council, 2021a; 2021b; 2021c), the mechanism provides grants, guarantees, and concessional loans for economic diversification, worker retraining, and clean-energy infrastructure in regions dependent on fossil-fuel industries. By linking social protection, regional investment, and low-carbon development, the Just Transition Mechanism operationalizes the distributive, restorative, and recognition dimensions of justice.

Complementing the Just Transition Mechanism, the Modernization Fund, financed through the EU ETS, supports 13 lower-income EU member states in modernizing their energy systems and improving energy efficiency, while prioritizing measures that protect vulnerable consumers and workers (European Parliament and Council, 2018b). In addition, several national compensatory schemes address the diverse social and regional impacts of energy transitions. These include coal-sector mechanisms providing financial support to utilities, municipalities, and affected workers to facilitate the managed decline of coal production while maintaining social cohesion (e.g., Germany, Poland); industrial transition funds aligned with the EU Just Transition Mechanism supporting reskilling and diversification in carbon-intensive regions (e.g., Spain, Czech Republic); energy price-compensation measures shielding consumers during periods of market volatility (e.g., France, Italy); and others.

Globally, compensatory and corrective mechanisms are now embedded within international climate-finance and development architectures. For example, the Loss and Damage Fund operationalized under the Paris Agreement advances restorative justice by channeling finance to countries incurring climate-related losses (United Nations, 2023). Complementing this, the Climate Investment Funds (the Clean Technology Fund and the Strategic Climate Fund) established by the World Bank integrate just-transition objectives across programs to mitigate distributional harms and support place-based redevelopment (World Bank, 2008). Just Energy Transition Partnerships are established as a political initiative under the United Nations Framework Convention on Climate Change mobilize blended finance for socially inclusive power-sector decarbonization (e.g., South Africa, Indonesia, and Vietnam), linking emissions reductions with worker and community safeguards (European Commission, 2022). The Global Environment Facility established by the World Bank finances remediation and resilience co-benefits associated with phasing out carbon-intensive activities (World Bank, 1994). In parallel, environmental remediation and reskilling funds—grounded in the World Bank's "Just Transition for All" framework and the International Labour Organization's Just Transition Guidelines—provide compensation, land and asset repurposing, and workforce transition support in coal- and fossil-fuel-dependent regions (International Labour Organization, 2015). Together, such instruments extend compensatory and corrective justice to the global scale, aligning climate action with energy-sector equity, employment protection, and community regeneration.

Despite their growing prominence, compensatory and corrective mechanisms face persistent challenges related to governance, equity targeting, and policy coherence. In some cases, compensation has been criticized for favoring incumbent industries or reinforcing regional dependency on state transfers, rather than fostering structural transformation. Effective compensation therefore requires transparent eligibility criteria, social dialogue, and participatory planning to ensure that benefits reach affected workers and communities rather than corporate actors. Furthermore, corrective measures must be closely coordinated with redistributive, market-based, and investment instruments to form a coherent equity architecture that balances prevention, mitigation, and restoration.

It can be stated that compensatory and corrective mechanisms provide the remedial dimension of equity governance, ensuring that transitions toward cleaner and more resilient energy systems do not exacerbate existing inequalities or create new forms of exclusion. By addressing the social and territorial consequences of structural change, these instruments reinforce the legitimacy and durability of decarbonization pathways.

All global and regional mechanisms can also be conceptualized as international financial cooperation instruments. These instruments embody a shift from nationally bounded conceptions of justice toward a cosmopolitan perspective, recognizing that fairness in the global energy transition must extend beyond domestic equity concerns. In this view, responsibilities for decarbonization, adaptation, and social protection are shared collectively across nations, reflecting the moral and economic interdependence of an integrated energy and climate system.

Economic and financial instruments, in all their forms, thus provide the tangible means through which equity commitments are realized across the energy sector. When designed in accordance with justice principles, they redistribute resources, correct historical imbalances, and enhance participation in energy decision-making. Their effectiveness ultimately depends on policy coherence, institutional capacity, and sustained political will.

7.2.3 Technology, Innovation, and Digital Inclusion

Technological innovation and digitalization represent transformative forces in contemporary energy systems, reshaping production, consumption, and governance processes. Their impacts on social equity are profound and multidimensional. On the one hand, innovation can reduce costs, enhance efficiency, and expand access to modern energy services; on the other, it can reinforce social divisions if technological and digital benefits are unevenly distributed. The pace and direction of technological change therefore determine whether innovation functions as an equalizing or differentiating force within the energy transition.

In principle, technology and digitalization can strengthen distributional equity by lowering barriers to energy access and participation; enhance procedural equity by enabling wider stakeholder involvement through data and transparency tools; and support recognition equity by tailoring solutions to diverse social and territorial contexts. However, empirical evidence demonstrates that innovation outcomes are contingent on policy design, institutional capacity, and social readiness. Without targeted interventions, the advantages of smart and digital technologies tend to accrue disproportionately to high-income, urban, and digitally skilled groups, while low-income or older consumers remain excluded from efficiency gains and decision-making.

7.2.3.1 Technological Innovation and Equity Outcomes

Technological progress contributes to social equity through several pathways. First, distributed and renewable energy technologies—such as rooftop solar, community wind, or microgrids—enable decentralized access, foster local ownership, and strengthen community resilience. When supported by equitable financing and capacity-building, these systems can reduce both economic and spatial disparities. Second, digital technologies such as smart meters, demand-response platforms, and home-energy-management systems empower consumers to monitor and adjust their energy use, improving affordability and autonomy. Nevertheless, these opportunities remain unevenly accessible, particularly where digital infrastructure, financial resources, or technical competence are limited.

This digital-innovation divide produces differentiated capacities to benefit from low-carbon transitions. High-income or technologically advanced users may become "energy prosumers" participating in local markets, while others remain passive consumers, bearing residual system costs. Such disparities highlight the need for equity-oriented innovation governance that explicitly integrates inclusion, affordability, and accessibility into technological design, regulation, and implementation.

Technological and digital readiness constitutes a key determinant of both innovation acceptance and the distribution of its associated benefits. It is shaped by multiple socio-demographic factors—including age, socioeconomic determinants, social norms, or gender—that collectively influence individuals' capacity to access, adopt, and effectively utilize emerging technologies (Jones et al., 2015; Shimoda et al., 2020; Schwartz and Krarti, 2022; Zhang, 2023; Wali et al., 2025). For example, empirical research shows that women and men engage with emerging household and energy technologies through distinct social roles and competencies shaped by divisions of care and domestic labor (Hargreaves et al., 2018; Strengers et al., 2022). Smart-home and digital-energy systems often reflect masculine assumptions of autonomy, control, and technical expertise, reducing usability and adoption among women. These asymmetries stem not from lower interest but from exclusionary design practices and unequal time availability. Addressing them requires targeted interventions—gender-inclusive design standards, participatory technology testing, and digital-literacy programs—ensuring that innovation contributes equitably to social inclusion rather than reinforcing existing disparities.

Empirical studies show that older adults often possess lower digital competence, shorter investment horizons, and more established routines, limiting their engagement with smart or automated energy systems (Szulecki et al., 2024). These readiness deficits hinder effective uptake of technologies such as heat pumps, solar installations, demand-response platforms, or home-energy-management systems by older cohorts, resulting in a concentration of benefits among younger, more digitally skilled consumers. At a broader structural level, aging populations also pose significant challenges for the development and diffusion of green innovation. Liu and Jia (2023) find that in China, population aging exerted a significant inhibitory effect on green innovation efficiency. Consequently, innovation advantages risk bypassing segments of the population defined by age unless policymakers deploy targeted interventions—including age-tailored digital-literacy programs, technology design adapted to older users, and investment support with shorter payoff horizons—to ensure that emerging energy systems contribute to rather than undermine social equity.

Education also constitutes a critical determinant of technological readiness and the capacity to engage with energy innovations. Zhou et al.'s (2023) findings showed that higher levels of educational attainment significantly enhance renewable-energy demand, confirming the role of education in strengthening both cognitive and behavioral readiness for low-carbon technologies. Educated populations are better equipped to interpret environmental information, utilize digital tools, and make informed energy choices, thereby accelerating the diffusion of green technologies. Conversely, limited education constrains individuals' awareness, technical competence, and trust in innovation, reducing participation in renewable-energy adoption and digitalized energy-management systems. These results highlight that education does not only facilitate economic growth but also underpins social inclusion in technological transitions. In addition, evidence from Meng and Hao (2024) confirms that expanding higher education contributes positively to sustainable energy development, with each 1% growth in the higher-educated population yielding a 0.19% short-term and a 0.29% long-term improvement in sustainability outcomes. Accordingly, integrating education-responsive measures such as lifelong learning, public information campaigns, and capacity-building for digital and environmental literacy into energy policy design is essential to ensure that the benefits of energy innovation are distributed equitably and contribute to sustainable transformation.

Beyond educational attainment, other socioeconomic and demographic determinants—particularly income and expenditure capacity, household composition, market access, and spatial location—systematically shape who adopts and benefits from energy innovations. Recent household-level evidence demonstrates that household income and expenditure capacity strongly influence the adoption of solar home systems, confirming that affordability remains a decisive factor in renewable energy uptake (Wali et al., 2025). Schwartz and

Krarti (2022) identify socioeconomic conditions, including income, housing characteristics, and ownership structures, as key correlates of adoption across energy-efficiency and distributed-energy technologies in the U.S. residential sector. Similarly, Min (2023) highlights that social and spatial inequalities affect the distribution of rooftop-solar and electric-vehicle infrastructure in urban areas, with higher-income communities exhibiting markedly higher adoption rates and technological diffusion. Collectively, these studies underscore that technological innovation is not adopted uniformly across society: individuals and communities with greater economic resources and social capital are more likely to access and benefit from emerging energy technologies, while economically constrained groups risk exclusion from innovation gains.

7.2.3.2 Digital Inclusion As a Dimension of Social Equity

Digital inclusion has become a central dimension of fairness within contemporary energy systems. It extends beyond mere physical access to digital infrastructure—such as broadband connectivity, smart meters, and digital devices—to encompass the financial capacity, cognitive ability, and trust required to use data-driven energy services effectively. Globally, digital inequality remains a defining challenge of the low-carbon transition. In many developing and emerging economies, insufficient connectivity, affordability barriers, and low digital literacy constrain access to smart technologies and limit participation in efficiency or demand-response programs. Initiatives such as the ITU's Partner2Connect Digital Coalition, UNDP's Digital Strategy 2.0, and the World Bank's Digital Economy for Africa programme aim to bridge these divides by expanding connectivity, enhancing digital skills, and fostering inclusive innovation ecosystems (ITU, 2022; World Bank, 2025; UNDP, 2022). However, these efforts remain less integrated and sector-specific than those of the developed economies. At the same time, several advanced economies—such as the United States, Canada, Japan, and Australia—have introduced digital equity and connectivity programs that link broadband access, digital literacy, and smart-energy infrastructure as part of broader national innovation agendas. For example, in the context of the EU, the Digital Economy and Society Index (DESI) and the Digital Decade Policy Programme 2030 (European Parliament and Council, 2022) reveal persistent digital divides across income groups, regions, genders, and age cohorts, which translate directly into differentiated participation in smart-energy markets, dynamic pricing schemes, and efficiency programs.

Digital literacy thus emerges as a prerequisite for equitable participation in digitalized energy systems, ensuring that individuals and communities can effectively engage with and benefit from technological innovation. In this context, digital inclusion represents a new layer of distributional equity, as digital access now determines who can share in the economic and environmental benefits of the energy transition. Households without broadband connectivity or smart devices are systematically excluded from dynamic tariffs, energy-efficiency incentives, and peer-to-peer energy trading. This creates a "digital divide in decarbonization," where economic and environmental advantages accrue disproportionately to digitally connected and higher-income groups, reinforcing structural inequalities. For example, empirical evidence from Yue et al. (2024) shows that the digital divide hampers efforts to alleviate energy poverty, reinforcing a Matthew effect whereby digitally excluded households become increasingly disadvantaged. This effect is particularly pronounced in low-income countries and rural areas, where limited digital infrastructure exacerbates vulnerability. Also, it was concluded that the digital divide aggravates energy poverty mainly by widening income inequality, thereby undermining the distributive goals of the energy transition.

At the same time, digital inclusion advances procedural and recognition equity by broadening participation in decision-making and acknowledging diverse user needs. Access to digital platforms allows citizens and communities to engage in energy governance—through data transparency, participatory monitoring, and smart-community initiatives—strengthening procedural justice. Furthermore, recognizing the specific digital vulnerabilities of groups such as elderly persons, women, rural residents, and persons with disabilities ensures that technological transitions respect social diversity rather than reproduce exclusionary patterns. Hence, integrating digital inclusion into energy policy is not merely a technical or connectivity issue but a core equity imperative, essential for achieving an inclusive and just energy transition that aligns with Sustainable Development Goals 7, 9, and 10 (United Nations, 2015b).

7.2.3.3 Innovation Governance and Inclusive Technology Policy

The governance of innovation plays a decisive role in shaping equitable outcomes. Inclusive research and innovation frameworks, emphasizing social value, participation, and diversity, help align technological development with justice principles. For instance, the European Commission's Horizon Europe Strategic Plan (European Commission, 2024) promotes citizen engagement, gender balance, and regional participation in energy research and innovation. Likewise, frugal innovation initiatives in emerging economies adapt technological solutions to local contexts and affordability constraints, thereby broadening recognition of diverse user needs (Pansera and Owen, 2018).

Social-innovation approaches further advance procedural equity by involving communities in the co-creation and testing of new solutions. Community-energy laboratories and living labs, for example, demonstrate how participatory experimentation can democratize technology development while strengthening local capabilities (Trenks and Bögel, 2024). These initiatives demonstrate that the fairness of technological transitions depends not only on how technologies are designed but also on how they are governed—through participatory processes, inclusive rules, and equitable distribution of benefits.

Beyond research and experimentation, innovation governance also encompasses the regulatory and market frameworks that determine who can participate in and benefit from technological change. Embedding equity into technology policy therefore requires a comprehensive approach that addresses affordability, accessibility, and accountability. This includes establishing inclusive standards and procurement rules to ensure universal usability of digital-energy systems; defining transparent, nondiscriminatory access conditions for smart grids and energy data; and introducing consumer-protection and data-governance mechanisms that safeguard privacy, consent, and fair algorithmic treatment. To be effective, national innovation policies should incorporate systematic equity-impact assessments that evaluate the social and distributional implications of emerging technologies before large-scale deployment.

It is necessary to stress that transparency and accountability mechanisms are essential for maintaining public trust in technology-driven transitions. Continuous monitoring through equity-sensitive indicators, including disaggregated adoption rates, affordability impacts, and user satisfaction, supports adaptive governance that responds to disparities as they emerge. Integrating such monitoring into national energy and innovation strategies ensures that digitalization and technological progress serve not only efficiency and decarbonization objectives but also fairness and inclusiveness. In this way, governance that prioritizes inclusion and equity can transform technological progress into a genuine force for social justice and a pathway toward sustainable, inclusive energy systems.

7.2.3.4 Integrating Technology and Digital Inclusion into Equitable Governance

Integrating technology and digital inclusion into equitable energy governance requires a deliberate policy approach that combines innovation, social protection, and participatory governance. Equity cannot emerge as a byproduct of digitalization; it must be embedded from the outset in how technologies are designed, financed, and regulated. Policymakers therefore need to align innovation strategies with social objectives, ensuring that the deployment of digital and low-carbon technologies supports affordability, accessibility, and empowerment for all social groups.

A key step is to establish governance frameworks that treat digital access as a social right, comparable to energy access itself. This involves integrating broadband and data infrastructure into national energy and social policies, particularly in rural and peripheral regions where digital exclusion coincides with energy vulnerability.

Equitable governance also depends on cross-sectoral coordination between ministries and agencies responsible for energy, social policy, education, and digital development. Digital inclusion and energy equity are interdependent challenges that require integrated policy responses—linking, for instance, social-tariff design with smart-meter rollout or energy-efficiency retrofits with digital-literacy initiatives.

At the same time, participatory and transparent decision-making processes are essential to sustain legitimacy in technology-driven transitions. Citizens, communities, and civil-society organizations should be meaningfully involved in shaping digital-energy frameworks—through public consultations, community-energy councils, or participatory budgeting mechanisms. These approaches not only strengthen procedural equity but also enhance trust in innovation, ensuring that digital transformation remains socially responsive and democratically accountable.

Finally, integrating technology and digital inclusion into equitable governance requires the development of monitoring and evaluation frameworks capable of tracking both technological performance and social impact. Indicators should capture who benefits from digitalization, who is excluded, and how costs and advantages are distributed across regions and population groups. Transparency in data use, algorithmic accountability, and open access to disaggregated statistics enable continuous learning and policy adjustment.

Table 7.5 Governance domains for equitable technological and digital transitions. Source: developed by the author.

Governance domain	*Governance objectives*	*Governance mechanisms/ illustrative initiatives*	*Expected social outcomes*
Innovation design and research frameworks	Embed inclusiveness, participation, and diversity into research and innovation agendas.	Horizon Europe; frugal and social innovation programs.	Technologies adapted to user diversity; improved legitimacy and social acceptance of innovation.
Regulatory and market frameworks	Define transparent rules for access, pricing, and consumer protection in smart and digital-energy systems.	Inclusive standards and procurement rules; nondiscriminatory grid and data-access regulation; privacy-by-design and algorithmic-fairness requirements.	Fair participation in digital-energy markets; reduced exclusion from innovation benefits.
Cross-sectoral coordination	Align energy, social, education, and digital policies to ensure coherent equity outcomes.	Interministerial coordination bodies; equity observatories; linkage of smart-meter rollout with social-tariff and literacy programs.	Policy coherence across sectors; balanced territorial and social outcomes.
Participation and transparency	Strengthen citizen involvement and accountability in technology governance.	Community-energy councils; living labs; participatory budgeting and consultation platforms.	Increased public trust, empowerment, and democratic legitimacy of innovation.
Monitoring and evaluation frameworks	Track technological performance together with social and distributional effects.	Equity-impact assessments; disaggregated indicators on access, affordability, and user satisfaction; open data on inclusion metrics.	Evidence-based adaptive governance; early identification and correction of disparities.
Integration of digital inclusion	Treat digital access and literacy as social rights integral to energy citizenship.	Broadband-expansion programs, digital-literacy initiatives, inclusive public–private partnerships.	Narrowed digital divide; improved participation of vulnerable and rural populations.

From a governance perspective, the integration of technology and digital inclusion into equitable energy systems can be organized across several interrelated domains. Each domain performs specific functions—from innovation design and regulatory oversight to participatory engagement and social-impact monitoring—collectively shaping how fairness is embedded in technological change. As summarized in Table 7.5, these domains demonstrate that equity depends as much on institutional design and governance coherence as on technological advancement.

Collectively, these governance domains demonstrate that achieving social equity in technology-driven energy transitions requires more than the deployment of innovative solutions—it necessitates a deliberate alignment between technological progress and institutional design. Effective governance should embed digital inclusion, transparency, and participation across all policy levels, ensuring that innovation functions not only as a driver of efficiency but also as a tool for social empowerment and inclusiveness.

7.3 Integrating Analytical and Policy Pathways: Toward Inclusive Energy Governance

Advancing social equity in the energy sector requires bridging analytical frameworks with practical policy implementation. The preceding sections have demonstrated that equitable energy governance emerges through the interaction of measurement methodologies, regulatory and economic instruments, and technology-driven innovation. Yet, ensuring coherence across these dimensions demands continuous learning, institutional coordination, and international cooperation. Energy transitions are inherently global processes: technological change, market dynamics, and social inequalities transcend national borders, making policy transfer and cross-national collaboration indispensable for achieving equity at scale.

International cooperation and policy learning therefore play a pivotal role in aligning national equity strategies with global sustainability objectives. Platforms such as the United Nations Framework Convention on Climate Change (UNFCCC), the International Energy Agency (IEA), and the International Renewable Energy Agency (IRENA) facilitate knowledge exchange, data harmonization, and best-practice dissemination on just-transition policies, energy-poverty alleviation, and inclusive innovation. Similarly, regional frameworks—such as the EU's Just Transition Mechanism, ASEAN's Power Integration initiatives, and the African Union's Agenda 2063 (African Union Commission, 2015)—illustrate how transnational policy coordination strengthens social and territorial equity. By translating international experience into domestic governance and fostering mutual learning among policymakers, such cooperation enhances the adaptive capacity of institutions and supports equitable implementation across different socioeconomic contexts.

In this context, integrating analytical and policy pathways implies more than the technical alignment of indicators and instruments; it represents a strategic process of embedding equity across governance layers, from local implementation to global collaboration. Therefore, inclusive energy governance depends on the ability of institutions to translate empirical evidence into actionable policy and to apply lessons from international experience to context-specific challenges.

A central prerequisite for such integration is the establishment of analytical coherence between measurement tools and decision-making processes. Equity assessments, whether based on multidimensional indices, affordability metrics, or distributive analyses, should directly inform regulatory design, fiscal allocation, and program evaluation. To achieve this, data collection must be harmonized across institutions and spatial scales, enabling policymakers to link empirical findings on inequality, energy access, and vulnerability with policy interventions in real time. The development of interoperable data infrastructures, open-access platforms, and shared indicator frameworks can significantly improve policy responsiveness and transparency.

Equally important is institutional learning, understood as the continuous adaptation of governance mechanisms through feedback, monitoring, and cross-sectoral exchange. Equity outcomes depend not only on initial policy design but also on the capacity of governments to adjust instruments as social and technological conditions evolve. The creation of a regular evaluation system, in which insights from program monitoring, stakeholder feedback, and comparative analysis are systematically incorporated into policy adjustments, establishes a dynamic governance process that enhances both effectiveness and legitimacy.

Moreover, multi-level coordination is crucial for ensuring that equity principles are consistently implemented from local to international levels. Local authorities play an essential role in tailoring national frameworks to territorial and social realities, particularly in addressing energy poverty, access disparities, and digital exclusion. National governments, in turn, must provide enabling regulatory and financial environments, while international bodies facilitate policy diffusion, capacity-building, and solidarity mechanisms. Aligning these scales transforms equity from a normative commitment into an operational governance principle.

Finally, achieving inclusive energy governance requires the institutionalization of equity mainstreaming—the systematic integration of fairness considerations into all stages of energy policy design, implementation, and evaluation. This involves embedding equity and justice dimensions within strategic planning, impact assessments, and investment criteria. When combined with robust analytical evidence, participatory mechanisms, and international policy learning, such mainstreaming ensures that equity becomes an inherent component of energy transition governance rather than an ex-post corrective measure.

The evidence-based and socially equitable energy governance framework illustrates how analytical evidence, participatory implementation, and institutional learning interact to create inclusive and adaptive energy governance (Fig. 7.2). It captures the essential linkages between knowledge generation,

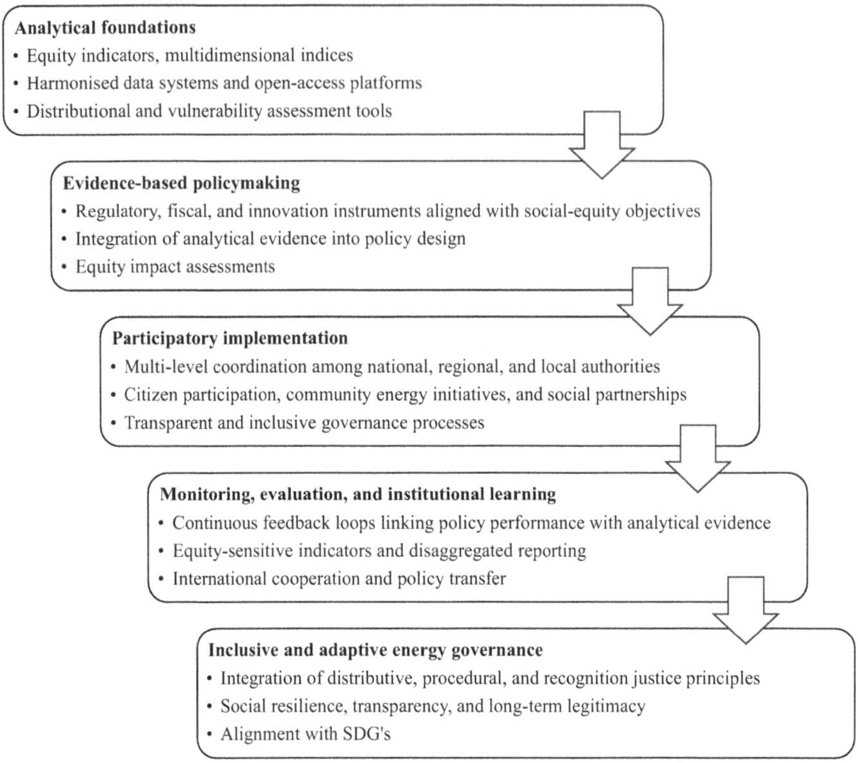

Fig. 7.2 Framework for evidence-based and socially equitable energy governance. Source: developed by the author.

policymaking, and stakeholder participation that collectively embed equity within the governance of energy transitions. Analytical foundations provide empirical insights into fairness and social impacts; evidence-based policymaking translates these insights into regulations and investment priorities; participatory implementation enhances legitimacy and responsiveness; and continuous monitoring and cross-national learning sustain policy relevance over time. Together, these elements depict how evidence and practice converge to support an equitable and socially grounded transition toward sustainable energy systems.

Although the framework is presented in a structured and sequential form for clarity, in practice these relationships should function as a continuous and iterative process. Equitable governance evolves through reciprocal exchange between analyses, policymaking, and implementation—each stage informing and refining the others. Analytical work must be revisited as new evidence and technologies emerge; policy instruments should adapt to lessons from implementation; and participatory feedback should guide institutional learning and future policy design. This iterative dynamic underscores that equity-oriented governance cannot be static: it requires continuous adjustment to changing social needs, technological advances, and contextual realities.

Integrating analytical and policy pathways therefore transforms energy governance into a learning-oriented and adaptive system. By linking empirical analysis with decision-making, fostering cross-sectoral collaboration, and drawing on global experience, policymakers can ensure that energy systems evolve not only toward decarbonization and efficiency but also toward fairness, inclusiveness, and long-term social resilience.

References

African Union Commission. Agenda 2063: The Africa we want. 2015, 24 p. Available at: https://au.int/sites/default/files/documents/36204-doc-agenda2063_popular_version_en.pdf

Asian Development Bank. Energy Transition Mechanism. 2025. Available at: https://www.adb.org/what-we-do/energy-transition-mechanism-etm

Baker, S.; Broadbent, A.; Bullock, C.; Heffron, R. Metrics for decision-making in energy justice. Annual Review of Environment and Resources, 2023, 48, pp. 589–611.

Barlow, J.; Tapio, R.; Tarekegne, B. Advancing the state of energy equity metrics. The Electricity Journal, 2022, 35(10), 107208.

Bartiaux, F.; Vandeschrick, C.; Moezzi, M.; Frogneux, N. Energy justice, unequal access to affordable warmth, and capability deprivation: a quantitative analysis for Belgium. Applied Energy, 2018, 225, pp. 1219–1233.

Bhatia, M.; Angelou, N. Beyond Connections: Energy Access Redefined (ESMAP Technical Report 008/15). Washington, DC: World Bank, 2015, p. 228. Available at: https://openknowledge.worldbank.org/server/api/core/bitstreams/248a7205-e926-5946-9025-605b8035ad95/content

Boardman, B. Fuel Poverty: From Cold Homes to Affordable Warmth. Belhaven Press, London, 1991, p. 267. ISBN 1852931396.

Cong, S.C.; Nock, D.; Qiu, Y.L.; Xing, B. Unveiling hidden energy poverty using the energy equity gap. Nature Communications, 2022, 13, 2456.

European Commission. EU Emissions Trading System. 2025b. Available at: https://climate.ec.europa.eu/eu-action/carbon-markets/eu-emissions-trading-system-eu-ets_en

European Commission. Horizon Europe Strategic Plan 2025–2027. Brussels, 2024, p. 140. doi: 10.2777/092911

European Commission. Joint Statement: South Africa Just Energy Transition Investment Plan. Brussels, 2022, p. 4. Available at: https://ec.europa.eu/commission/presscorner/api/files/document/print/en/statement_22_6664/STATEMENT_22_6664_EN.pdf

European Commission. National energy and climate plans. 2025a. Available at: https://commission.europa.eu/energy-climate-change-environment/implementation-eu-countries/energy-and-climate-governance-and-reporting/national-energy-and-climate-plans_en

European Commission. Regional Policy: New Cohesion Policy. 2021. Available at: https://ec.europa.eu/regional_policy/2021-2027_en

European Commission. Regulatory learning in the EU: Guidance on regulatory sandboxes, testbeds, and living labs in the EU, with a focus section on energy. Brussels,

29.8.2023 SWD(2023) 277/2 final. 2023, Available at: https://research-and-innovation.ec.europa.eu/system/files/2023-08/swd_2023_277_f1.pdf

European Commission. The European Green Deal (COM(2019) 640 final). Brussels, 2019. Available at: https://eur-lex.europa.eu/legal-content/EN/TXT/?uri=CELEX%3A52019DC0640

European Parliament; Council. Decision (EU) 2022/2481 of the European Parliament and of the Council of 14 December 2022 establishing the Digital Decade Policy Programme 2030. Official Journal of the European Union, 2022, L 323/4, pp. 4–26. Available at: https://eur-lex.europa.eu/eli/dec/2022/2481/oj

European Parliament; Council. Directive (EU) 2018/2001 of the European Parliament and of the Council of 11 December 2018 on the promotion of the use of energy from renewable sources. Official Journal of the European Union, 2018a, L 328, pp. 82–209. Available at: https://eur-lex.europa.eu/eli/dir/2018/2001/oj/eng

European Parliament; Council. Directive (EU) 2018/410 of the European Parliament and of the Council of 14 March 2018 amending Directive 2003/87/EC to enhance cost-effective emission reductions and low-carbon investments, and Decision (EU) 2015/1814. Official Journal of the European Union, 2018b, L 76, pp. 3–27. Available at: https://eur-lex.europa.eu/legal-content/EN/TXT/?uri=CELEX%3A32018L0410

European Parliament; Council. Directive (EU) 2019/944 of the European Parliament and of the Council of 5 June 2019 on common rules for the internal market for electricity and amending Directive 2012/27/EU. Official Journal of the European Union, 2019, L 158, pp. 125–199. Available at: https://eur-lex.europa.eu/eli/dir/2019/944/oj/eng

European Parliament; Council. Directive 2003/87/EC of the European Parliament and of the Council of 13 October 2003 establishing a scheme for greenhouse gas emission allowance trading within the Community and amending Council Directive 96/61/EC. Official Journal of the European Union, 2003, L 275, pp. 32–46. Available at: https://eur-lex.europa.eu/legal-content/EN/TXT/?uri=CELEX%3A32003L0087

European Parliament; Council. Regulation (EU) 2021/1056 of the European Parliament and of the Council of 24 June 2021 establishing the Just Transition Fund. Official Journal of the European Union, 2021a, L 231, pp. 1–20. Available at: https://eur-lex.europa.eu/legal-content/EN/TXT/?uri=CELEX:32021R1056

European Parliament; Council. Regulation (EU) 2021/1058 on the European Regional Development Fund and on the Cohesion Fund. Official Journal of the European Union, 2021d, L 231, pp. 60–93. Available at: https://eur-lex.europa.eu/legal-content/EN/TXT/?uri=CELEX%3A32021R1058

European Parliament; Council. Regulation (EU) 2021/1229 of the European Parliament and of the Council of 14 July 2021 on the public sector loan facility under the Just Transition Mechanism. Official Journal of the European Union, 2021c, L 274, pp. 1–19. Available at: https://eur-lex.europa.eu/eli/reg/2021/1229/oj/eng

European Parliament; Council. Regulation (EU) 2021/523 of the European Parliament and of the Council of 24 March 2021 establishing the InvestEU Programme and amending Regulation (EU) 2015/1017. Official Journal of the European Union, 2021b, L 107, pp. 30–89. Available at: https://eur-lex.europa.eu/eli/reg/2021/523/oj/eng

European Parliament; Council. Regulation (EU) 2023/955 establishing the Social Climate Fund and amending Regulation (EU) 2021/1060. Official Journal of the European Union, 2023, L 130, pp. 1–51. Available at: https://eur-lex.europa.eu/eli/reg/2023/955/oj/eng

European Parliament; Council; Commission. Interinstitutional Proclamation on the European Pillar of Social Rights. Official Journal of the European Union, 2017, C 428, pp. 10–15. Available at: https://eur-lex.europa.eu/legal-content/EN/TXT/?uri=oj:JOC_2017_428_R_0009

Eurostat. European Union Statistics on Income and Living Conditions (EU-SILC). 2025. Available at: https://ec.europa.eu/eurostat/web/main/data/database?node_code=livcon

Feenstra, M.; Özerol, G. Energy justice as a search light for gender-energy nexus: Towards a conceptual framework. Renewable and Sustainable Energy Reviews, 2021, 138, 110668.

Hargreaves, T., Wilson, C., Hauxwell-Baldwin, R. Learning to live in a smart home. Building Research & Information, 2018, 46(1), 127–139.

Heindl, P.; Schuessler, R. Dynamic properties of energy affordability measures. Energy Policy, 2015, 86, p. 123–132.

Hills, J. Fuel Poverty: The problem and its measurement. Interim Report of the Fuel Poverty Review, 2011, Centre for Analysis and Social Exclusion, Case Report 69, p. 192, ISSN 1465-3001.

Hills, J. Getting the measure of fuel poverty: Final report of the fuel poverty review. Case Report 72, 2012, p. 237, ISSN 1465-3001.

Hoßfeld, T.; Skorin-Kapov, L.; Heegaard, P.E.; Varela, M. Definition of QoE fairness in shared systems. IEEE Communication Letters, 2016, 21(1), pp. 184–187.

Huang, Y.; Barker, T. The Clean Development Mechanism and low carbon development: A panel data analysis. Energy Economics, 2012, 34(4), pp. 1033–1040.

IEA, IRENA, UNSD, World Bank, WHO. Tracking SDG 7: The Energy Progress Report. World Bank, Washington DC, 2025, p. 177. Available at: https://trackingsdg7.esmap.org/sites/default/files/download-documents/sdg7-report2025-0804-v11.pdf

International Energy Agency. World energy outlook 2024. International Energy Agency, 2024, p. 396. Available at: https://iea.blob.core.windows.net/assets/140a0470-5b90-4922-a0e9-838b3ac6918c/WorldEnergyOutlook2024.pdf

International Labour Organization. Guidelines for a just transition towards environmentally sustainable economies and societies for all. Switzerland, 2015, p. 23. 978-92-2-130628-3. Available at: https://www.ilo.org/sites/default/files/wcmsp5/groups/public/%40ed_emp/%40emp_ent/documents/publication/wcms_432859.pdf

ITU. Partner2Connect Action Framework Report. 2022, p. 156. Available at: https://www.itu.int/itu-d/reports/partner2connect-action-framework/wp-content/uploads/sites/7/2022/03/Partner2Connect-Action-Framework-Report.pdf

Jacobson, A.; Milman, A.D.; Kammen, D.M. Letting the (energy) Gini out of the bottle: Lorenz curves of cumulative electricity consumption and Gini coefficients as metrics of energy distribution and equity. Energy Policy, 2005, 33 (14), pp. 1825–1832.

Jenkins , K .; McCauley , D .; Heffron , R .; Stephan , H .; Rehner , R . Energy justice: A conceptual review. Energy Research & Social Science, 2016, 11, 174–182.

Jones, R.V.; Fuertes, A.; Lomas, K.J. The socio-economic, dwelling and appliance related factors affecting electricity consumption in domestic buildings. Renewable and Sustainable Energy Reviews, 2015, 43, pp. 901–917.

Jové-Llopis, E.; Trapero-Bertran, M.; Trujillo-Baute, E. Energy poverty and health: does the social energy tariff help? Energy Policy, 2025, 206, 114762.

Karpavicius. T.; Balezentis T. Public Service Obligation Levy in the Context of Energy Sustainability and Security: The Cases of Ireland, Greece, Denmark and Lithuania. Energies, 2022, 15(1), 16.

Liu, Y.; Jia, M. The Impact of Population Aging on Green Innovation: An Empirical Analysis Based on Inter-Provincial Data in China. Sustainability, 2023, 15, 3305.

Marte, M.; van Dender, K. The use of revenues from carbon pricing. OECD Taxation Working Papers No. 43. OECD, 2019. 68 p. Available at: https://www.sipotra.it/wp-content/uploads/2019/06/The-use-of-revenues-from-carbon-pricing.pdf

Meng, B.; Hao, Z. Role of green finance and higher education in fostering the sustainability and energy transition practices. Humanities and Social Sciences Communications, 2024, 11, 1298.

Min, Y. Clean energy justice: Different adoption characteristics of rooftop solar and EV chargers in Seattle. Energy Research & Social Science, 2023, 97, 102995.

Moore, R. Definitions of fuel poverty: implications for policy, Energy Policy, 2012, 49, pp. 19–26.

Moret, F.; Pinson, P. Energy collectives: A community and fairness based approach to future electricity markets. IEEE Transactions on Power Systems, 2019, 34(5), pp. 3994–4004.

Nussbaumer, P.; Bazilian, M.; Modi, V. Measuring energy poverty: Focusing on what matters. Renewable Sustainable Energy Review, 2012, 16, pp. 231–243.

Okedele , P . O .; Aziza , O . R .; Oduro , P .; Ishola , A . O . Carbon pricing mechanisms and their global efficacy in reducing emissions: Lessons from leading economies. Open Access Research Journal of Engineering and Technology, 2024, 7(2), 114–125

Onolemhemhena, R.; Bello, S. Does carbon pricing policy improve energy efficiency? New evidence from Canadian provinces. International Journal of Green Energy, 2024, 21(13), pp. 3072–3083.

Pansera, M.; Owen, R. Framing inclusive innovation within the discourse of development: Insights from case studies in India. Research Policy, 2018, 47(1), pp. 23–34.

Ren, W.H.; Guan, Y.P.; Qiu, F.; Levin, T.; Heleno, M. Energy justice and equity: A Review of definitions, measures, and practice in policy, planning, and operations. Renewable and Sustainable Energy Reviews, 2025, 222, 115900.

Schittekatte, T.; Meeus, L.; Jamasb, T.; Llorca, M. Regulatory experimentation in energy: Three pioneer countries and lessons for the green transition. Energy Policy, 2021, 156, 112382.

Schwartz, E.K.; Krarti, M. Review of Adoption Status of Sustainable Energy Technologies in the US Residential Building Sector. Energies, 2022, 15, 2027.

Shimoda, Y.; Yamaguchi, Y.; Iwafune, Y.; Hidaka, K.; Meier, A.; Yagita, Y.; Kawamoto, H.; Nishikiori, S. Energy demand science for a decarbonized society in the context of the residential sector. Renewable Sustainable Energy Review, 2020, 132, 110051.

Siksnelyte-Butkiene, I.; Streimikiene, D.; Lekavicius, V.; Balezentis, T. Energy poverty indicators: A systematic literature review and comprehensive analysis of integrity. Sustainable Cities and Society, 2021, 67, 102756.

Sovacool, B. K., Axsen, J., Sorrell, S.. Promoting novelty, rigor, and style in energy social science: Towards codes of practice for appropriate methods and research design. Energy Research & Social Science, 2019, 45, pp. 12–42.

Strengers, Y.; Gram-Hanssen, K.; Dahlgren, K.; Aagaard, L. K. Energy, emerging technologies and gender in homes. Buildings and Cities, 2022, 3(1), pp. 842–853.

Szulecki, K., Neerland, M.A., Tomter, H. et al. Ageism, welfare, and the energy transition: a comparative analysis of the perceptions among the elderly in Poland and Norway. Energy, Sustainability and Society, 2024, 14, p. 35.

Tarekegne, B. W., Pennell, G. R., Preziuso, D. C.; O'Neil, R. S. (2021). Review of Energy Equity Metrics. Pacific Northwest National Laboratory, Richland, Washington, 25 p. Available at: https://doi.org/10.2172/1830804

Theil, H. Economics and information theory. Amsterdam: North Holland, 1967.

Theil, H. World income inequality and its components. Economics Letters, 1979, 2(1), pp. 99–102.

Trenks, H.; Bögel, P.M. Empowering citizens for the energy transition: facilitating role change through real-world experiments. Sustainability Science, 2024, 19, pp. 715–737.

UNDP. Digital Strategy 2022–2025. 2022, p. 52. Available at: https://digitalstrategy.undp.org/documents/Digital-Strategy-2022-2025-Full-Document_ENG_Interactive.pdf

United Nations. Framework Convention on Climate Change. Decision 3/CP.17: Launching the Green Climate Fund. FCCC/CP/2011/9/Add.1, 2012, p. 86. Available at: https://unfccc.int/resource/docs/2011/cop17/eng/09a01.pdf#page=55

United Nations. Framework Convention on Climate Change. Decision 2/CMA.4: Funding arrangements for responding to loss and damage associated with the adverse effects of climate change, including a focus on addressing loss and damage. FCCC/PA/CMA/2022/10/Add.1. 2023. Available https://unfccc.int/sites/default/files/resource/cma2022_10_a01E.pdf#page=13

United Nations. Kyoto Protocol to the United Nations Framework Convention on Climate Change. 1998, p. 21. Available at: https://unfccc.int/resource/docs/convkp/kpeng.pdf

United Nations. Paris agreement. 2015b, p. 25. Available at: https://unfccc.int/sites/default/files/english_paris_agreement.pdf

United Nations. Transforming our world: The 2030 Agenda for Sustainable Development (A/RES/70/1). 2015a. p. 41. Available at: https://www.un.org/ohrlls/sites/www.un.org.ohrlls/files/2030_agenda_for_sustainable_development_web.pdf

Wali, H.N.; Dutta, T.; Sani, B. et al. Socioeconomic determinants of solar energy adoption and installed capacity in urban areas of Kano State, Nigeria. Discover Sustainability, 2025, 6, 1095.

World Bank. Instrument for the Establishment of the Restructured Global Environment Facility. Report of the GEF Participants Meeting. Geneva, Switzerland, March 14–16, 1994. Available at: https://documents1.worldbank.org/curated/en/282361492622558487/pdf/Instrument-for-the-establishment-of-the-restructured-Global-Environment-Facility-GEF.pdf

World Bank. Memorandum of the President of the Climate Investment Funds: on the Establishment the Clean Technology Fund and the Strategic Climate Fund. 2008, Washington, DC 20433. Available at: https://documents1.worldbank.org/curated/en/134611468138289814/txt/Climate-investment-funds-the-clean-technology-fund-and-the-strategic-climate-fund.txt

World Bank. RISE 2024. Regulatory Indicators for Sustainable Energy: The Time Is Now. 2024, p. 52. Available at: https://documents1.worldbank.org/curated/en/099051625184514754/pdf/Regulatory-Indicators-for-Sustainable-Energy-RISE-2024.pdf

World Bank. The Digital Economy for Africa Initiative. 2025. Available at: https://www.worldbank.org/en/programs/all-africa-digital-transformation

World Energy Council. World Energy Trilemma Index 2023. 2023. Available online at: https://trilemma.worldenergy.org/

Yitzhaki, S.; Schechtman, E. The Gini methodology: A primer on a statistical methodology. Springer. 2013, p. 548.

Yue, Q.; Zhang, M.; Song, Y. Impact of digital divide on energy poverty across the globe: The mediating role of income inequality. Energy Policy, 2024, 195, 114349.

Zhang, M. Older people's attitudes towards emerging technologies: A systematic literature review. Public Understanding of Science, 2023, 32(8), pp. 948–968.

Zhou, Y.; Liu, Z.; Wang, M.; Dong, R.K.; Yue, X.G. Evaluating the impacts of education and digitalization on renewable energy demand behaviour: new evidence from Japan. Economic Research-Ekonomska Istraživanja, 2023, 36(3) 2164033.

CHAPTER 8

Applying Social Equity Frameworks: An Illustrative Example of Age-Inclusive Low-Carbon Transitions

8.1 Policy and Governance Approaches

The preceding chapters of this book have explored the multiple dimensions of social equity in the energy sector, including energy justice, energy poverty, gender, and age. While these discussions provide a conceptual and analytical foundation, translating social equity frameworks into actionable policies requires concrete illustrations. This chapter presents an applied example, focusing specifically on age-inclusive approaches in low-carbon energy transitions.

Age has been selected as the illustrative lens because aging populations face distinctive vulnerabilities and barriers in accessing and benefiting from clean energy systems. Moreover, the concept of "energy ageism"—the structural and social disadvantages experienced by older adults in the energy transition—remains underexplored relative to other dimensions of equity. By examining age-inclusive policy and governance approaches, implementation challenges, and methodological considerations, this chapter demonstrates how broader social equity frameworks can be operationalized in practice.

This example serves both as a practical illustration of applying social equity frameworks and as a lens through which broader lessons for inclusive energy policy can be drawn, ensuring that the principles of justice and equity are embedded in real-world energy transitions.

8.1.1 *Just Transition and Welfare State Approaches*

8.1.1.1 Just Transition Frameworks and Age Inclusion

The concept of "just transition" has emerged as a central framework for ensuring that the shift to low-carbon energy systems addresses social equity

I. Siksnelyte-Butkiene, D. Streimikiene, *Social Equity in the Energy Sector*,
https://doi.org/10.1007/978-3-032-22224-4_8

concerns and does not leave vulnerable populations behind, recognizing that energy system transformations create both winners and losers and that deliberate policy interventions are necessary to ensure that the costs and benefits of low-carbon energy transitions are fairly distributed across different social groups and geographic areas (Creti and Ftiti, 2024). However, the integration of age considerations into just low-carbon energy transition frameworks has been limited, despite growing evidence that older adults face unique challenges and opportunities in the context of energy system change that require systematic analysis and targeted intervention beyond general approaches to social equity and inclusion.

In Table 8.1 just low-carbon energy transition frameworks with an age-inclusive lens are given.

Table 8.1 outlines a comprehensive framework for integrating age inclusion into just low-carbon energy transition policies by breaking the concept down into four key pillars: equitable distribution, procedural justice, recognition justice, and restorative and social investment. For each pillar, it first defines the general principle, then details its specific application to the challenges faced by older adults, such as protecting them from high costs and ensuring their participation in decision-making. The table further operationalizes these concepts by posing critical policy questions designed to guide the creation of fair and effective measures, while a final column provides the main academic references that underpin these frameworks.

Traditional just low-carbon energy transition frameworks have focused primarily on workers and communities affected by the decline of fossil-fuel industries, with particular attention to coal mining regions and communities dependent on fossil fuel-related employment that face economic disruption and social challenges as energy systems transition to cleaner alternatives (Goddard and Farrelly, 2018). While these concerns remain important and require continued attention, the emergence of energy ageism as a conceptual framework suggests that just low-carbon energy transition approaches need to be expanded to address a broader range of social groups and forms of disadvantage that may be affected by energy system transformations in ways that require different types of policy responses and support mechanisms.

The integration of age considerations into just low-carbon energy transition frameworks requires recognition that older adults may be affected by low-carbon energy transitions in ways that differ from other social groups and that require different types of policy responses and support mechanisms that address their specific needs, preferences, circumstances, and capabilities (Żuk and Żuk, 2024a,b). Unlike younger workers who may be able to retrain for new employment opportunities in clean energy industries, older adults may face different challenges related to adapting to new energy technologies, navigating changing energy markets, and accessing the benefits of clean energy investments while also having different resources and capabilities that can contribute to successful low-carbon energy transitions.

Table 8.1 Just transition frameworks with an age-inclusive lens. Source: created by the authors based on Healy and Barry (2017), Sovacool (2021), Walker (2012), Żuk and Żuk (2024a), Schlosberg (2007), Żuk and Żuk (2022), ILO (2015), Jenkins et al. (2016), and Tikkakoski et al. (2024).

Dimension	*Application to age inclusion*	*Key policy questions*	*References*
Equitable distribution	Shield seniors on fixed incomes from energy price shocks and the high upfront costs of green technologies. Ensure they directly benefit from energy efficiency through lower bills and improved comfort, rather than being left behind.	How can social tariffs or energy allowances protect pensioners? Are subsidies for home retrofits structured as grants that cover 100% of the cost for low-income seniors?	Healy and Barry, 2017; Sovacool, 2021
Procedural justice	Actively involve older adults in policy design by overcoming common barriers. This means providing non-digital options for consultation and support, using clear language, and offering direct, in-person assistance.	Are senior citizens' organizations formally included in policy consultations?Are support programs designed with offline, paper-based application routes and clear, simple instructions?	Walker, 2012; Żuk and Żuk, 2024a
Recognition justice	Recognize seniors as a distinct group with specific needs related to health, mobility, and fixed incomes. Policy must move beyond a one-size-fits-all approach and be tailored to their specific living conditions and capabilities.	Do policies explicitly recognize age as a key axis of vulnerability in the energy transition?Are the health impacts of cold homes on the elderly considered a primary driver for policy action?	Schlosberg, 2007; Żuk and Żuk, 2022
Restorative and social investment	Direct public funds to correct decades of neglect in housing. This involves systemic, state-funded thermal modernization of entire districts with high concentrations of older residents and ensuring health services are prepared for climate impacts like heatwaves.	Should public investment prioritize upgrading the housing stock of the most vulnerable seniors? How can social and health services be integrated with energy policy to protect older adults?	ILO, 2015; Jenkins et al. 2016; Tikkakoski et al., 2024.

Recent policy developments have begun to recognize the need for more inclusive approaches to just low-carbon energy transition that address age-related concerns alongside other dimensions of social equity, reflecting a growing understanding that low-carbon energy transitions affect multiple social groups in different ways and that comprehensive approaches to just transition must address this diversity (Heffron, 2022). As noted in recent energy policy analysis, there has been a realization that the energy sector was missing an overall raison d'être, and that is to have justice permeating throughout the sector. This recognition has led to increased attention to the need for low-carbon energy transitions that are "fair, equal, equitable and inclusive" across all dimensions of social difference, including age (Heffron, 2022).

The development of age-inclusive, just low-carbon energy transition frameworks requires attention to several key principles and mechanisms that can help ensure that older adults are not left behind in energy system transformations while also leveraging their capabilities and contributions to successful energy system change. It is necessary to stress that just low-carbon energy transition approaches must recognize the diversity of older adult experiences and avoid one-size-fits-all solutions that may not address the specific needs and circumstances of different groups of older adults (Jenkins et al., 2016). This requires disaggregated analysis of how low-carbon energy transitions affect older adults with different income levels, housing situations, health conditions, and geographic locations, as well as recognition that older adults are not a homogeneous group with identical income, needs, and preferences.

Also, age-inclusive, just low-carbon energy transition approaches must address both the barriers that older adults face in accessing the benefits of low-carbon energy transitions and the contributions that older adults can make to successful energy system transformation (Szulecki et al., 2024). This requires moving beyond deficit-based approaches that focus only on older adults' vulnerabilities to recognize their capabilities, knowledge, and resources that can contribute to more effective and sustainable energy transitions, including their experience with energy conservation, their knowledge of local conditions and community needs, and their potential contributions to community-energy initiatives and social sustainability.

Just low-carbon transition frameworks must include mechanisms for meaningful participation by older adults in energy transition planning and implementation, addressing the procedural justice concerns identified in energy ageism research and ensuring that public participation processes are accessible to older adults and that their perspectives and concerns are meaningfully incorporated into energy policy development (Fuller and McCauley, 2016). This requires not only ensuring that participation processes are physically and technologically accessible to older adults but also that they are designed to accommodate different participation styles and preferences and that they value the different types of knowledge and experience that older adults can contribute.

Overall, age-inclusive, just low-carbon energy transition approaches must address the temporal dimensions of low-carbon energy transitions that may create particular challenges for older adults, including the mismatch between the time horizons of energy investments and older adults' expected tenure in their current homes; the need for immediate benefits rather than long-term payoffs; and the importance of ensuring that low-carbon energy transitions do not create short-term disruptions that disproportionately affect older adults (Jenkins et al., 2018). This may require innovative financing mechanisms, accelerated implementation timelines for programs that benefit older adults, and careful attention to the sequencing of low-carbon energy transition policies to minimize disruptions.

8.1.1.2 Welfare State Models and Energy Ageism Mitigation

The comparative research on energy ageism in different national contexts (section 6.3.1) showed how welfare state institutions and social policy approaches can mitigate age-related barriers to low-carbon energy transition participation, with important implications for understanding how different policy models can promote or hinder age-inclusive low-carbon energy transitions (Szulecki et al., 2024). The contrast between Norway's social democratic welfare state model and Poland's more limited welfare state provides important lessons about the role of social policy in promoting age-just, low-carbon energy transitions, while also highlighting the limitations of welfare state approaches and the need for targeted interventions that address specific dimensions of energy ageism.

The Norwegian model demonstrates how comprehensive social protection systems can help address some of the economic barriers that contribute to energy ageism, with higher pension levels, universal healthcare, and comprehensive social services helping to ensure that older adults have the economic security necessary to invest in energy efficiency improvements and new energy technologies (Szulecki et al., 2024). The Norwegian model also includes more extensive public investment in energy infrastructure, which helps ensure that the benefits of clean energy transitions are more broadly distributed across the population rather than being concentrated among those with the resources to make individual investments in new technologies and services.

The Norwegian approach to energy policy also includes more extensive consultation and participation processes that may be better positioned to address older adults' concerns and preferences, while the country's strong tradition of social solidarity and intergenerational cooperation may provide a more supportive context for developing age-inclusive energy policies (Szulecki et al., 2024). The Norwegian welfare state model also includes more comprehensive approaches to housing policy and urban planning that may help address some of the spatial dimensions of energy ageism by ensuring that older adults have access to energy-efficient housing and energy services regardless of their economic circumstances.

However, the Norwegian case also demonstrates that robust welfare states alone are not sufficient to address all dimensions of energy ageism, with even Norwegian older adults facing significant challenges related to digital competence and technological barriers that suggest that targeted interventions are necessary to address specific age-related barriers to energy transition participation (Szulecki et al., 2024). This finding highlights the importance of combining comprehensive social protection with targeted interventions that address specific barriers to energy access and participation, while also suggesting that technological change in energy systems may create new forms of exclusion that require ongoing attention and intervention.

The Polish case illustrates how more limited welfare state institutions can exacerbate energy ageism by leaving older adults more vulnerable to energy poverty and less able to access the benefits of energy transitions (Szulecki et al.,

2024). Lower pension levels, more limited social services, and greater reliance on market-based approaches to energy provision create additional barriers for older adults seeking to participate in energy transitions, while also limiting their ability to cope with energy system changes that may disrupt traditional energy sources and practices.

The Polish case also highlights how historical legacies of energy system development can create additional challenges for low-carbon energy transitions, with older adults in Poland being more dependent on coal heating systems that are both environmentally harmful and economically vulnerable to price volatility and policy changes (Żuk and Żuk, 2024a). The transition away from coal heating requires significant investments in new heating systems and may involve disruptions to familiar practices and routines that can be particularly challenging for older adults who may have limited experience with alternative heating technologies.

In Table 8.2 the welfare state models and their approaches to mitigating energy ageism are provided.

Table 8.2 above provides a comparative analysis of how different welfare state models—social democratic, liberal, conservative, and post-socialist—are equipped to mitigate energy ageism, drawing on foundational typologies (Esping-Andersen, 1990). It systematically outlines each model's core principles, its likely approach to tackling energy ageism, and its inherent strengths and weaknesses in protecting older citizens during an energy transition. By placing these distinct political-economic frameworks side-by-side, the table illustrates how a country's broader commitment to social welfare, market intervention, and equality fundamentally shapes its ability to move from reactive, targeted support to the proactive, systemic solutions needed to effectively prevent energy poverty among its aging population.

These comparative insights suggest several important principles for designing welfare state approaches that can help mitigate energy ageism and promote age-inclusive, just low-carbon energy transitions. The comprehensive social protection systems that ensure adequate income security for older adults are essential for addressing the economic barriers that contribute to energy ageism (Szulecki et al., 2024). This includes not only pension systems but also healthcare, housing assistance, and other social services that affect older adults' overall economic security and their ability to invest in energy improvements and adapt to energy system changes.

Also, public investment in energy infrastructure and services can help ensure that the benefits of low-carbon energy transitions are more broadly distributed and that older adults are not excluded from access to clean energy technologies and services (Upham et al., 2022). This might include public investment in energy efficiency and retrofitting programs for older housing stock, subsidies for clean energy technologies, and public provision of energy services in areas where market provision is inadequate or inaccessible to older adults with limited economic resources.

Table 8.2 Welfare state models and their approaches to mitigating energy ageism. Source: created by the author based on Esping-Andersen (1990), Szulecki et al. (2024), Galvin (2020), Fenger (2007), and Żuk and Żuk (2024a).

Model	*Core principle*	*Approach to energy ageism mitigation*	*Strengths in mitigation*	*Potential weaknesses*	*Source*
Social democratic (Nordic)	Universalism and De-commodification: Aims to provide high-level social protection to all citizens as a social right, minimizing market reliance.	Proactive and Systemic: Mitigation is integrated into a broad, state-led system. This involves universal benefits, state-funded housing modernization, and strong public utilities that prioritize social welfare.	High Prevention: Universal benefits and robust pensions prevent many seniors from falling into energy poverty. Systemic Solutions: Focuses on structural fixes (e.g., upgrading all housing) rather than just individual support.	High Cost: Requires significant public expenditure and high levels of taxation. Bureaucratic: Can be slow to adapt to new challenges if not designed to be flexible.	Esping-Andersen, 1990; Szulecki et al. (2024)
Liberal (Anglo-Saxon)	Market Dominance and Individualism: Emphasizes a minimal social safety net, relying on the market to provide for individuals. State support is often means-tested and targeted only at the most needy.	Reactive and Targeted: Mitigation relies on means-tested support for the "deserving poor" (e.g., energy bill rebates or crisis payments). Housing upgrades are left to individual homeowners or the private market.	Lower Tax Burden: Less costly for the state in terms of direct public spending.Efficient Targeting (in theory): Directs limited resources to the most desperate cases.	High Stigma and Low Uptake: Means-testing can be complex and stigmatizing, leading many eligible seniors not to apply.• Creates Gaps: Fails to support those just above the poverty line and does not address root causes.	Esping-Andersen, 1990

(continued)

Table 8.2 (continued)

Model	*Core principle*	*Approach to energy ageism mitigation*	*Strengths in mitigation*	*Potential weaknesses*	*Source*
Conservative / Corporatist (Continental European)	Social Insurance and Familialism: Social protection is tied to an individual's employment history (social insurance). The family is also expected to be a primary provider of support.	Status-Preserving and Segmented: Mitigation support is often linked to past contributions (pensions). It may rely on a mix of state-regulated energy prices and an expectation that families will support older relatives.	Stability for Insiders: Provides decent protection for those with stable, lifelong employment histories.Social Partnerships: Can facilitate agreements between the state and energy companies to protect consumers.	Excludes Outsiders: Fails to protect those with fragmented careers or outside traditional family structures.Reinforces Inequality: The level of protection mirrors an individual's past position in the labor market.	Esping-Andersen, 1990
Post-Socialist (Central and Eastern European)	Hybrid and Transitional: A mix of inherited state structures, liberal market reforms, and familial reliance. Often characterized by underfunded social systems and low trust in state institutions.	Underdeveloped and Inconsistent: Mitigation is often ad hoc and poorly funded. It may consist of short-term subsidies or programs requiring co-payments that are unaffordable for many. There is a heavy reliance on family support.	Potential for State Action: The legacy of state ownership can provide a basis for direct intervention. High Awareness of Need: The scale of the problem is often well-understood by the public.	Lack of Resources and Trust: Insufficient public funds and low trust in government lead to ineffective programs. High Vulnerability: Low pensions and inefficient housing stock create a "perfect storm" for energy ageism.	Fenger, 2007; Żuk and Żuk 2024a

In addition, the targeted programs and services may be necessary to address specific age-related barriers to low-carbon energy transition participation, including digital-literacy training programs, accessible information and outreach services, and specialized support for older adults navigating energy system changes (Bouzarovski et al., 2023). These targeted interventions should be designed to complement rather than replace comprehensive social protection systems, while also being integrated with broader energy transition policies to ensure coherence and effectiveness.

Finally, welfare state approaches to addressing energy ageism must be sensitive to the diversity of older adult experiences and avoid one-size-fits-all solutions that may not address the specific needs and circumstances of different groups of older adults (Krekula et al., 2018). This requires attention to how age intersects with other forms of inequality based on income, race, gender, geography, and other factors, as well as recognition that older adults have different preferences, capabilities, and circumstances that require flexible and responsive policy approaches.

The framework in Table 8.3 presents a policy framework that integrates financial, infrastructural, and social support systems to ensure older adults are not left behind in the low-carbon energy transition.

Table 8.3 gives a comprehensive and actionable framework of policies and measures designed to ensure an age-just, low-carbon energy transition, systematically organizing interventions across four critical domains: financial support, infrastructural interventions, governance, and community support. It moves beyond abstract principles to outline specific, concrete actions that governments and institutions can take to protect older adults from the negative impacts of decarbonization. In the Financial Support category, it emphasizes direct, proactive measures like universal energy allowances and fully funded deep retrofit grants, which are designed to remove the economic barriers that are often the primary driver of energy ageism. The Infrastructural and Housing section shifts the focus from individual responsibility to systemic, state-led solutions, advocating for area-based modernization of entire housing districts and the expansion of public energy networks to address the root causes of energy inefficiency at scale. Recognizing that how policies are made is as important as what they do, the Governance and Procedural Fairness category details the necessity of inclusive program design with non-digital pathways and the formal inclusion of seniors in co-designing the policies that affect them, thereby ensuring relevance and building trust. Finally, the Health and Community Support section broadens the scope to include the crucial integration of energy policy with public health services and the empowerment of local communities through support for collective energy projects. By structuring the measures in a systematic way and supporting them by the main academic references, the table serves as a holistic roadmap for policymakers, illustrating that a truly just low-carbon energy transition for older adults requires a multipronged approach that simultaneously addresses income, housing, participation, and well-being.

Table 8.3 Policies and measures for an age-just, low-carbon energy transition. Source: created by the authors based on Walker and Day (2012), Healy and Barry (2017), Sovacool (2021), Żuk and Żuk (2024a), Bouzarovski et al. (2018), Tikkakoski et al. (2024), Sokołowski and Bouzarovski (2022), Jenkins et al. (2016), Willand et al. (2022), Walker (2012), Szulecki et al. (2024), Awaworyi Churchill and Smyth (2021), Curtis et al. (2018), and Stephens (2019).

Policy category	*Specific measure*	*Description*	*Desired outcome*	*References*
Financial Support and Social Protection	Direct Income Support	Implement noncontributory, universal energy allowances or social tariffs for low-income seniors to shield them from energy price volatility and ensure affordability of essential energy services.	Prevents energy poverty by directly supplementing income, ensuring seniors do not have to choose between heating, food, and medicine.	Walker and Day, 2012; Healy and Barry, 2017
	Deep Retrofit Grant Programs	Offer state-funded grants that cover 100% of the cost for deep thermal modernization (insulation, window replacement) and furnace replacement for low-income older homeowners, with no co-payment required.	Removes the primary financial barrier to home-energy efficiency, leading to warmer homes, lower bills, and reduced emissions without indebting seniors.	Sovacool, 2021 Żuk and Żuk, 2024a
Infrastructural and Housing Interventions	Systemic Housing Modernization	Move beyond individual grants to state-led, area-based initiatives that systematically upgrade entire districts of old, inefficient housing stock, particularly those with high concentrations of older residents.	Addresses the root cause of energy inefficiency at scale, improving living conditions for entire communities and avoiding a piecemeal approach.	Bouzarovski et al. 2018; Tikkakoski et al., 2024
	Expansion of Public Energy Infrastructure	Invest in and expand access to municipal district heating and gas networks in underserved urban and rural areas, providing a cleaner and often more stable alternative to solid fuels.	Provides a structural solution that reduces reliance on polluting fuels, improves local air quality, and offers a more convenient and often more affordable heating source.	Sokołowski and Bouzarovski, 2022
Governance and Procedural Fairness	Inclusive and Accessible Program Design	Design support programs with simple, clear language and provide multiple non-digital application channels (e.g., in-person at community centers, paper forms). Offer one-stop-shop advisory services to guide seniors through the process.	Increases uptake of support programs by removing bureaucratic and digital barriers, ensuring that aid reaches those who need it most.	Jenkins et al., 2016; Willand et al., 2022

	Formal Consultation and Co-Design	Establish formal, institutionalized roles for senior citizens' representative organizations in the co-design and evaluation of all energy transition policies and programs.	Ensures that policies are grounded in the lived reality of older adults, making them more effective, trusted, and just. It shifts seniors from being passive recipients to active participants.	Walker, 2012; Szulecki et al., 2024
Health and Community Support	Integration of Health and Energy Policy	Formally link energy poverty mitigation with public health goals. This includes training healthcare and social care workers to identify signs of energy deprivation and refer seniors to support services.	Creates a proactive public health response to the negative health impacts of cold homes, preventing illness and reducing strain on the healthcare system.	Awaworyi Churchill and Smyth, 2021; Curtis et al. 2018
	Support for Community-Energy Initiatives	Provide funding and technical assistance for community-owned renewable energy projects (e.g., solar panels on cooperative housing blocks), with schemes that allow older residents to directly benefit from lower energy costs.	Empowers communities, fosters social cohesion, and provides a decentralized, resilient energy source that can directly reduce energy bills for participating seniors.	Stephens, 2019

8.1.2 Governance, Policy Instruments, and Institutional Mechanisms

8.1.2.1 Multi-level Governance and Policy Coordination

The complexity of energy ageism and the multiple dimensions through which it operates require coordinated policy responses across multiple levels of government and multiple policy domains, with effective responses requiring coordination between national energy policy, regional planning, local housing policy, and social services (Żuk and Żuk, 2024 a, b). The research on spatial determinants of energy ageism demonstrates that addressing age-related energy disadvantages requires understanding how different levels of governance interact to create or mitigate barriers to energy access and participation, while also highlighting the importance of coordinated approaches that address multiple scales and dimensions simultaneously.

At the national level, energy policy frameworks need to explicitly consider age-related impacts and include mechanisms for ensuring that low-carbon energy transitions do not exacerbate age-based inequalities while also leveraging older adults' capabilities and contributions to energy system transformation (Heffron, 2022). This might include age impact assessments for major energy policy initiatives, requirements for age-inclusive design of energy programs, and targeted funding for addressing age-related barriers to low-carbon energy transition participation. National energy policies should also include provisions for monitoring and evaluating the age-related impacts of low-carbon energy transitions to ensure that energy and climate policy goals are being achieved and to identify areas where additional interventions may be needed.

National social policy frameworks also play crucial roles in addressing energy ageism by providing the broader social protection systems that help mitigate economic vulnerability and ensure that older adults have the resources necessary to participate in low-carbon energy transitions (Szulecki et al., 2024). This includes pension systems, healthcare programs, housing assistance, and other social services that affect older adults' overall well-being and economic security. The integration of energy considerations into social policy frameworks can help ensure that social programs support rather than hinder low-carbon energy transition goals, while also ensuring that energy policies consider their impacts on older adults' access to social services and support systems.

National policy frameworks should also address the regulatory and institutional factors that may create barriers to age-inclusive low-carbon energy transitions, including utility regulations that may not adequately protect older adults from energy disconnections or price volatility, building codes and standards that may not address the needs of aging-in-place, and consumer-protection regulations that may not adequately address the specific vulnerabilities that older adults face in energy markets (Jenkins et al., 2016). These regulatory frameworks should be designed to promote age inclusion while also supporting broader low-carbon energy transition goals and ensuring that energy markets operate fairly and efficiently.

At the regional level, energy planning and infrastructure development need to consider the spatial dimensions of energy ageism and ensure that energy infrastructure investments address the needs of areas with high concentrations of older adults (Żuk and Żuk, 2024a). This might include prioritizing energy efficiency retrofits in neighborhoods with older housing stock, ensuring that renewable energy installations are accessible to older adults, and coordinating transportation and energy planning to address mobility-related barriers to energy access and participation.

Regional planning processes should also include mechanisms for older adult participation and should consider how regional development patterns affect older adults' access to energy services and their ability to participate in low-carbon energy transitions (Upham et al., 2022). This might involve regional aging councils or other mechanisms for ensuring that older adult perspectives are included in regional energy planning processes, as well as coordination between regional energy planning and regional aging services to ensure that energy and aging policies are mutually supportive.

Regional governance structures may also be important for coordinating between different local jurisdictions and ensuring that age-inclusive energy policies are implemented consistently across different communities and municipalities within a region. This is particularly important in metropolitan areas where older adults may live in different jurisdictions but share common energy infrastructure and services, requiring coordinated approaches that address regional rather than purely local factors.

Local-level policies and programs are particularly important for addressing the community and social capital dimensions of energy ageism, with local governments often being best positioned to understand the specific needs and circumstances of older adult populations in their communities and to develop targeted programs and services that address local barriers to low-carbon energy transition participation (McCauley et al., 2013). Local energy programs should be designed to be accessible to older adults and should include mechanisms for older adult's participation in program design and implementation while also being coordinated with local aging services and other social programs.

Local governments may also be important for addressing the spatial dimensions of energy ageism through zoning and land-use policies that promote age-friendly development patterns; building codes and standards that support aging-in-place; and housing policies that ensure that older adults have access to energy-efficient and affordable housing (Żuk and Żuk, 2024a). Local governments may also be important for implementing energy efficiency retrofit programs, renewable energy installations, and other energy transition initiatives that can benefit older adults while also contributing to broader climate change mitigation goals.

It is necessary to stress the importance of coordination across policy domains, as energy ageism intersects with housing policy, social services, healthcare, transportation, and other areas of public policy (Żuk and Żuk, 2024a, b). Effective responses to energy ageism require breaking down policy silos and developing integrated approaches that address the multiple dimensions through which age-related barriers operate. This might involve inter-agency coordination mechanisms, integrated service delivery models, and comprehensive planning processes that consider the interactions between different policy domains and their cumulative impacts on older adults' energy experiences.

8.1.2.2 Regulatory Frameworks and Consumer Protection

The development of age-inclusive low-carbon energy transitions also requires attention to regulatory frameworks and consumer-protection mechanisms that can help ensure that older adults are not disadvantaged by energy market changes and that they have access to appropriate protections and support when navigating increasingly complex energy systems (Bouzarovski et al., 2023). Traditional utility regulation has focused primarily on ensuring reliable service and reasonable rates for all consumers, but the increasing complexity, inclusion of AI, and digitization of energy systems may require new approaches to regulation that explicitly consider the needs and vulnerabilities of different consumer groups, including older adults.

Energy market deregulation and the introduction of competitive energy markets have created new opportunities for consumer choice and innovation, but they have also created new risks and challenges for consumers who may not have the knowledge, skills, or resources necessary to navigate complex market choices effectively (Furszyfer Del Rio et al., 2021). Older adults may be particularly vulnerable to predatory energy marketing practices, confusing contract terms, and other market failures that can result in higher energy costs or reduced service quality, requiring specific consumer-protection measures that address these vulnerabilities.

Regulatory frameworks should include specific protections for older adults and other vulnerable consumers, including restrictions on door-to-door marketing, requirements for clear and accessible contract terms, and protections against disconnection for nonpayment that consider older adults' specific circumstances and vulnerabilities (Jenkins et al., 2016). These protections should be designed to promote consumer choice and energy market competition while also ensuring that older adults are not disadvantaged by energy market changes or predatory practices.

The increasing digitization of energy systems also creates new regulatory challenges related to data privacy, algorithmic decision-making, and digital accessibility that may disproportionately affect older adults (Furszyfer Del Rio et al., 2021). Regulatory frameworks should include requirements for age-friendly design of digital-energy services, protections for older adults' personal data, and transparency requirements for algorithmic decision-making that affects energy service delivery. These requirements should be designed to promote innovation and efficiency while also ensuring that older adults can access and benefit from digital-energy services.

Regulatory frameworks should also address the accessibility and affordability of energy services for older adults, including requirements for utilities to offer accessible customer service options, flexible payment arrangements, and energy efficiency programs that are designed to meet older adults' specific needs and circumstances (Heffron, 2022). These requirements should be integrated with broader low-carbon energy transition policies to ensure that older adults can participate in and benefit from clean energy programs and services.

8.1.2.3 *Financing Mechanisms and Economic Instruments*

The development of age-inclusive low-carbon energy transitions requires innovative financing mechanisms and economic instruments that can help older adults access the upfront capital needed for energy investments while also addressing ongoing affordability concerns and ensuring that low-carbon energy transition benefits are distributed fairly across different age groups (Goddard and Farrelly, 2018). Traditional financing mechanisms for energy investments, such as loans or leasing arrangements, may not be accessible to older adults who may have limited credit history, fixed low incomes, or concerns about taking on long-term financial commitments that extend beyond their expected tenure in their current homes.

Innovative financing mechanisms that are specifically designed to meet older adults' needs and circumstances may include grant programs that provide upfront capital for energy investments without requiring repayment; on-bill financing programs that allow energy improvements to be paid for through utility bills over time; and community-based financing mechanisms that pool resources to make energy investments more affordable and accessible for all community members (Upham et al., 2022). These financing mechanisms should be designed to address the specific barriers that older adults face while also promoting broader low-carbon energy transition goals and ensuring that public resources are used effectively and efficiently.

Property assessed clean energy (PACE) financing programs, which allow property owners to finance energy improvements through property tax assessments, may be particularly beneficial for older adults because they transfer the financing obligation to future property owners if the property is sold before the financing is repaid (Goddard and Farrelly, 2018). However, PACE programs may also create challenges for older adults who may not fully understand the long-term implications of property tax assessments or who may face difficulties selling their properties if buyers are concerned about existing PACE obligations.

Targeted grant and rebate programs that provide direct financial assistance for energy efficiency improvements and renewable energy installations may be particularly important for older adults with limited incomes who cannot access traditional financing mechanisms (Żuk and Żuk, 2024a; Żuk and Żuk, 2024b). These programs should be designed to be accessible to older adults and should include appropriate outreach and education components to ensure that older adults are aware of available opportunities and can navigate application processes effectively.

Economic instruments such as progressive rate structures, lifeline rates, and targeted energy assistance programs can also help address affordability concerns for older adults while promoting energy conservation and efficiency (Heffron, 2022). These instruments should be designed to provide appropriate support for older adults with limited incomes while also creating incentives for energy conservation and efficiency that support broader energy transition goals.

8.1.2.4 Innovation and Technology Policy

The development of age-inclusive low-carbon energy transitions also requires attention to innovation and technology policy that can promote the development of energy technologies and services that are accessible to and beneficial for older adults (Bouzarovski et al., 2023). Traditional approaches to energy innovation have focused primarily on technical performance and cost-effectiveness, but age-inclusive innovation requires additional attention to usability, accessibility, and user experience factors that may be particularly important for older adults.

Innovation policy should include requirements or incentives for age-inclusive design of energy technologies and services, including consideration of factors such as ease of use, accessibility for people with disabilities, and compatibility with existing technologies and practices that older adults may be familiar with (Furszyfer Del Rio et al., 2021). This might include funding for research and development of age-friendly energy technologies, requirements for user testing with older adult populations, and standards for accessibility and usability that apply to energy technologies and services.

Technology policy should also address the digital divide and ensure that older adults can access and benefit from digital-energy services and technologies (Bouzarovski et al., 2023). This might include funding for digital-literacy programs, requirements for non-digital alternatives to digital-energy services, and support for the development of simplified and accessible digital interfaces that are designed specifically for older adult users.

Innovation policy should also promote the development of community-based and social-innovation approaches to low-carbon energy transitions that may be particularly beneficial for older adults, including community-energy programs, peer-to-peer learning initiatives, and social enterprises that focus on addressing energy-related challenges for older adults (McCauley et al., 2013). These approaches may be particularly important for addressing the social and community dimensions of energy ageism that cannot be addressed through technological solutions alone.

8.1.2.5 International Cooperation and Policy Learning

The development of age-inclusive low-carbon energy transitions can also benefit from international cooperation and policy learning that allows different countries and regions to share experiences, best practices, and innovative approaches to addressing energy ageism (Jenkins et al., 2016). International cooperation can help accelerate the development of effective policies and programs by allowing policymakers to learn from successes and failures in other contexts while also promoting the development of common standards and approaches that can facilitate technology transfer and knowledge sharing.

International organizations such as the International Energy Agency, the United Nations, and the European Union can play important roles in promoting age-inclusive low-carbon energy transitions by developing guidelines and standards for age inclusion in energy policy, facilitating knowledge sharing and policy learning between different countries and regions, and providing technical assistance and capacity building support for countries that are developing age-inclusive energy policies (Heffron, 2022).

International cooperation can also help address the global dimensions of low-carbon energy transitions and ensure that age inclusion is considered in international climate and energy agreements and initiatives (Creti and Ftiti, 2024). This is particularly important given that population aging is a global phenomenon that affects countries at different stages of development and with different energy system characteristics, requiring coordinated approaches that can address diverse contexts and circumstances.

Policy learning and knowledge sharing can also help identify innovative approaches and best practices that can be adapted to different contexts, while also highlighting common challenges and barriers that require coordinated responses (Upham et al., 2022). This might include comparative research on different policy approaches on age-inclusive low-carbon energy transitions and collaborative research and development programs that bring together researchers and practitioners from different countries and regions to solve this important problem.

8.2 Implementation Challenges of Age-just Low-carbon Transitions

Achieving an age-just, low-carbon energy transition requires overcoming a range of significant barriers that span economic, social, institutional, and personal domains. These obstacles often interact, creating complex challenges for older adults in low-carbon energy transitions. Table 8.4 summarizes the key barriers to an age-just, low-carbon energy transition.

Table 8.4 provides a comprehensive and multidimensional analysis of the key barriers that impede an age-just, low-carbon energy transition, systematically categorizing them into four interconnected domains: Economic and Financial, Institutional and Policy, Infrastructural and Spatial, and Social and Behavioral. By breaking down the challenges in this way, the table illustrates that the obstacles faced by older adults are not monolithic but are instead a complex web of mutually reinforcing issues. The Economic and Financial category highlights the most immediate and formidable barriers, such as the high upfront costs of green technologies and the fundamental inadequacy of low or fixed pensions, which trap seniors in a state of perpetual energy poverty. Building on this, the Institutional and Policy section reveals how systemic failures—including poorly designed subsidy programs with high bureaucratic hurdles and a lack of integration between energy, housing, and health policies—exacerbate these financial struggles and ensure that support often fails to reach those who need it

Table 8.4 The barriers of age-just, low-carbon energy transition. Source: created by the author based on Sovacool (2021), Axon and Morrissey (2020), Karpinska and Śmiech (2023), Żuk and Żuk (2024a), Matczak et al. (2023), Frankowski and Tirado Herrero (2021), Tikkakoski et al. (2024), Willand et al. (2022), Bouzarovski et al. (2018), Sokołowski and Bouzarovski (2022), Matczak et al. (2023), Żuk and Żuk (2023), and Szulecki et al. (2024).

Barrier category	*Specific barrier*	*Description*	*Impacts on older adults*	*Source*
Economic and Financial	High upfront costs of technology	The significant initial investment required for energy efficiency measures (e.g., insulation, new windows) and low-carbon technologies (e.g., heat pumps, solar panels) is prohibitive for those on low or fixed incomes.	Seniors on pensions often lack the savings or access to credit needed for these investments, trapping them in inefficient, high-cost energy systems.	Sovacool, 2021; Axon and Morrissey, 2020
	Inadequate income and low pensions	The primary determinant of energy poverty. Low and often fixed pensions do not keep pace with rising energy prices, leaving little to no disposable income for anything beyond basic necessities.	This directly leads to energy rationing (under-heating), fuel poverty, and the inability to absorb any additional costs associated with the transition, making it a source of constant financial stress.	Karpinska and Śmiech, 2023; Żuk and Żuk, 2024a
Institutional and Policy	Poorly designed subsidy programs	Support programs often require co-payments, reimbursement after the fact, complex paperwork, and digital-only applications, creating insurmountable hurdles for the most vulnerable.	These design flaws effectively exclude low-income seniors, those with low digital literacy, and those without savings, meaning the aid fails to reach its intended recipients.	Matczak et al., 2023; Frankowski and Tirado Herrero, 2021
Infrastructural and Spatial	Lack of policy integration	Energy policy is often developed in a silo, disconnected from social welfare, housing, and public health policies. This leads to a fragmented approach that fails to address the interconnected nature of the problem.	A senior might receive a new furnace but still live in a poorly insulated home, or their health issues from living in a cold home are not recognized as an energy-related problem by the healthcare system.	Tikkakoski et al., 2024; Willand et al., 2022
	Inefficient housing stock	Many older adults live in old, poorly insulated buildings (e.g., pre-war tenements and post-socialist blocks) that are difficult and expensive to heat, regardless of the energy source.	This is a structural barrier that locks seniors into high energy consumption and costs. Individual actions are insufficient without major, systemic retrofitting of the building stock.	Bouzarovski et al., 2018; Sokołowski and Bouzarovski, 2022

	Lack of access to modern infrastructure	Many older urban districts and rural areas lack access to cleaner energy infrastructure like district heating or gas networks, forcing a continued reliance on solid fuels like coal and wood.	This spatial inequality limits the energy choices available to seniors, often leaving them with only the most polluting, labor-intensive, and increasingly expensive options.	Matczak et al., 2023
Social and Behavioral	Low trust and risk aversion	A history of inconsistent policies and a lack of reliable state support can lead to deep-seated distrust in government programs and new technologies. Older adults may be risk-averse due to their financial precarity.	This can manifest as a reluctance to participate in subsidy programs or a "backup" mentality, such as keeping an old coal furnace "just in case" the new technology fails or becomes too expensive.	Frankowski and Tirado Herrero, 2021; Żuk and Żuk, 2023
	Information gaps and digital exclusion	Key information about the energy transition, available support, and new technologies is often communicated through digital channels and uses technical jargon that is inaccessible to many seniors.	This creates a significant information barrier, leaving older adults unaware of their options or unable to navigate the process, reinforcing feelings of exclusion and helplessness.	Szulecki et al., 2024

most. the infrastructural and spatial dimension further grounds these issues in the spatial dimension, showing how the legacy of inefficient housing stock and a lack of access to modern energy networks create structural traps that cannot be overcome by individual action alone. Finally, the Social and Behavioral category delves into the human element, detailing how low trust in government, risk aversion born from financial insecurity, and significant information gaps caused by digital exclusion create powerful psychological and practical barriers to participation. By presenting these distinct yet overlapping obstacles and supporting them with main academic references, the table serves as a critical diagnostic tool, making it clear that achieving an age-just, low-carbon energy transition requires a holistic strategy that simultaneously addresses income, policy design, physical infrastructure, and social trust.

8.2.1 Institutional and Organizational Barriers

The implementation of age-just, low-carbon energy transitions faces significant institutional and organizational barriers that reflect broader patterns of energy ageism within energy institutions and the limited integration of aging considerations into energy policy and planning processes (Jenkins, 2018). These barriers operate at multiple levels, from individual organizations to broader institutional frameworks that govern energy systems, and require systematic efforts to address both explicit and implicit forms of age-based discrimination that may be embedded in organizational cultures, policies, and practices.

At the organizational level, energy utilities, government agencies, and other energy institutions may lack the knowledge, capacity, or commitment necessary to address age-related barriers to low-carbon energy transition participation (Furszyfer Del Rio et al., 2021). Energy organizations may have limited understanding of older adults' needs and circumstances, may lack staff with expertise in aging issues, or may have organizational cultures that do not prioritize age inclusion in service delivery and program design. These organizational limitations may be particularly pronounced in technical organizations that focus primarily on engineering and economic considerations without adequate attention to social and demographic factors.

The studies show that many energy organizations operate with implicit assumptions about their customers and stakeholders that may not adequately reflect the diversity of older adult experiences (Bouzarovski et al., 2023; Heffron, 2022). Energy program design may assume technological proficiency, mobility, or economic resources that not all older adults possess, while customer service systems may not accommodate age-related changes in communication preferences or information processing needs. These implicit assumptions may not reflect intentional discrimination but can create systematic barriers that exclude older adults from accessing energy services and programs.

Addressing these organizational barriers requires systematic efforts to build capacity for age-inclusive service delivery within energy organizations, including training programs for energy professionals on aging issues and age-inclusive service delivery, recruitment and retention of staff with expertise in aging, development of age-friendly policies and procedures, and creation of feedback mechanisms that allow older adults to influence organizational practices (Heffron, 2022). These capacity-building efforts should be integrated with broader organizational development initiatives and should be supported by leadership commitment and adequate resources from state and municipalities.

Organizational change efforts should also address the cultural and attitudinal factors that may contribute to age-based exclusion, including ageist stereotypes and assumptions that may influence how energy professionals interact with older adults and design programs and services (Ayalon and Tesch-Römer, 2018). This might involve awareness training, psychological knowledge, policy development, and performance evaluation systems that promote age inclusion and hold staff accountable for providing age-friendly services.

At the broader institutional level, energy policy frameworks often lack explicit consideration of age-related impacts or mechanisms for ensuring age-inclusive implementation (Jenkins et al., 2016). Energy planning processes may not include systematic analysis of how proposed policies and programs will affect different age groups, may not include meaningful opportunities for older adult participation, or may not include accountability mechanisms for ensuring age-inclusive outcomes. These institutional limitations reflect broader patterns of age-based exclusion from policy processes that may require systematic reforms to address effectively.

Institutional barriers may also include legal and regulatory frameworks that do not adequately protect older adults from energy-related discrimination or that do not provide appropriate mechanisms for addressing age-related energy concerns (Bouzarovski et al., 2023). For example, utility regulations may not include specific protections for older adults, consumer-protection laws may not address age-related vulnerabilities in energy markets, and housing codes may not include requirements for age-friendly energy systems.

Addressing these institutional barriers requires systematic reforms that integrate age considerations into energy policy frameworks, regulatory systems, and planning processes (Heffron, 2022). This might involve legislative changes that require age impact assessments for energy policies, regulatory reforms that include specific protections for older adults, and planning process reforms that ensure meaningful older adult participation in energy decision-making.

8.2.2 *Digital Divides and Technological Barriers*

The increasing digitization of energy systems creates significant implementation challenges for age-just low-carbon transitions, with digital technologies having the potential to improve energy system efficiency and provide new opportunities for consumer engagement while also creating new barriers for older adults who lack digital access, skills, or confidence (Furszyfer Del Rio et al., 2021). The digital barriers operate at multiple levels and affect different aspects of energy system participation in ways that require comprehensive responses addressing both individual and systemic factors.

At the most basic level, some older adults lack access to reliable internet service or appropriate devices for accessing digital-energy services, with this digital divide being particularly pronounced in rural areas and among older adults with lower incomes (Bouzarovski et al., 2023). Even among older adults who have basic digital access, many lack the specific digital skills needed to navigate increasingly complex energy-management systems, including smart meters, home-energy-management platforms, and mobile applications for energy monitoring and optimization.

The digital divide in energy systems is not simply a matter of access to technology but also reflects broader issues of digital literacy, confidence, and trust that may be particularly challenging for older adults who may have had limited exposure to digital technologies or who may have concerns about privacy, security, and reliability (Szulecki et al., 2024). These concerns may be particularly pronounced for energy-related applications that involve personal data, financial transactions, or control of home systems that are critical for comfort and safety.

The research also reveals that digital barriers are not simply technical but reflect broader issues of design and usability that can be addressed through better design practices and more inclusive technology development (Furszyfer Del Rio et al., 2021). Many digital-energy platforms are designed with assumptions about user capabilities and preferences that may not reflect older adults' needs and circumstances, with complex interfaces, small text, and navigation systems that require multiple steps creating barriers even for older adults who have basic digital skills.

Addressing these technological and digital barriers requires both individual-level interventions and system-level changes that address the full range of factors that affect digital inclusion (Bouzarovski et al., 2023). Individual-level interventions might include digital-literacy training programs that focus specifically on energy-related applications, peer support programs that help older adults build confidence with digital technologies, and technical assistance services that provide ongoing support for navigating digital-energy systems.

System-level changes might include requirements for age-friendly design of digital-energy platforms, maintenance of non-digital alternatives for accessing energy services, and development of simplified interfaces and navigation systems that accommodate diverse user needs and capabilities (Heffron, 2022). The successful digital inclusion strategies must address both the supply side of digital services through improved design and accessibility and the demand side through education, support, and capacity-building programs.

The implementation of digital inclusion strategies also faces challenges related to the rapid pace of technological change in energy systems, which may outpace efforts to address digital barriers and may create new forms of exclusion even as previous barriers are addressed (Furszyfer Del Rio et al., 2021). This suggests the need for ongoing attention to digital inclusion and adaptive approaches that can respond to changing technological landscapes while maintaining focus on age inclusion goals.

8.2.3 Financial and Economic Constraints

The implementation of age-just low-carbon transitions also faces significant financial and economic constraints that reflect both the limited resources available to many older adults and the broader challenges of financing energy transitions in ways that are accessible to all population groups (Żuk and Żuk, 2024a). These financial constraints operate at multiple levels and require innovative financing mechanisms and policy interventions that can help older adults access the upfront capital needed for energy investments while also addressing ongoing affordability concerns.

Many older adults live on fixed incomes that may not provide sufficient resources for investing in energy efficiency improvements or renewable energy technologies, even when these investments could provide long-term cost savings (Szulecki et al., 2024). Traditional financing mechanisms for energy investments, such as loans or leasing arrangements, may not be accessible to older adults who may have limited credit history or who may be reluctant to take on long-term financial commitments that extend beyond their expected tenure in their current homes.

The temporal mismatch between energy-investment requirements and older adults' planning horizons creates particular challenges for financing age-inclusive energy transitions (Jenkins et al., 2018). Energy efficiency improvements and renewable energy installations typically require significant upfront investments that provide benefits over many years, but older adults may not expect to remain in their current homes long enough to recoup these investments through energy savings, creating a disincentive for participation in energy transition programs.

Financial constraints also affect the ability of energy organizations and government agencies to develop and implement age-inclusive energy programs, with limited budgets often leading to program designs that prioritize efficiency over accessibility (Upham et al., 2022). Addressing these constraints requires dedicated funding for age-inclusive energy programs and recognition that investing in age inclusion may require additional resources in the short term but can provide long-term benefits in terms of program effectiveness and social equity.

Addressing financial barriers to low-carbon energy transition participation requires developing innovative financing mechanisms that are specifically designed to meet older adults' needs and circumstances (Goddard and Farrelly, 2018). This might include grant programs that provide upfront capital for energy investments, on-bill financing programs that allow energy improvements to be paid for through utility bills, and community-based financing mechanisms that pool resources to make energy investments more affordable.

However, the implementation of innovative financing mechanisms also faces challenges related to regulatory barriers, administrative complexity, and the need for coordination between different organizations and funding sources (Heffron, 2022). These implementation challenges require systematic attention to program design, administration, and coordination to ensure that innovative financing mechanisms are accessible to older adults and effective in promoting energy transition participation.

8.2.4 Political and Social Resistance

The implementation of age-just, low-carbon energy transitions may also face political and social resistance from various sources, including older adults themselves who may be skeptical of low-carbon energy transition policies, younger generations who may prioritize environmental goals over age-related concerns, and political actors who may not see age inclusion as a priority (Fuller and McCauley, 2016). Understanding and addressing these sources of resistance is crucial for developing effective strategies for promoting age-just energy transitions that can build broad social support and political sustainability.

Some older adults may be resistant to low-carbon energy transition policies due to concerns about costs, disruptions to familiar routines, or skepticism about the benefits of new technologies (Szulecki et al., 2024). This resistance may be particularly pronounced among older adults who have experienced previous policy changes that negatively affected their well-being or who have limited trust in government institutions and energy companies. Resistance may also reflect concerns about the pace of change and the potential for energy transitions to disrupt established practices and relationships that are important for older adults' sense of security and well-being.

Addressing resistance from older adults requires understanding the sources of their concerns and developing communication and engagement strategies that address these concerns while also highlighting the potential benefits of low-carbon energy transitions (McCauley et al., 2013). This might involve peer-to-peer education programs, demonstration projects that allow older adults to see the benefits of new technologies, and policy designs that minimize disruptions and provide adequate support for transitions. Effective engagement strategies should also recognize and respect older adults' knowledge and experience while providing accurate information about low-carbon energy transition opportunities and benefits.

Political resistance to age-inclusive energy policies may come from various sources, including concerns about costs, competing priorities, and ideological opposition to government intervention in energy markets (Jenkins et al., 2016). Building political support for age-just, low-carbon energy transitions requires developing coalitions that include older adult advocacy organizations, environmental groups, and other stakeholders who can make the case for age inclusion in energy policy while also demonstrating the broader benefits of age-inclusive approaches for energy system effectiveness and social cohesion.

Political resistance may also reflect broader patterns of ageism in political discourse and policymaking that prioritize the interests of younger generations or that view older adults as obstacles to progress rather than as valuable contributors to social and environmental goals (Walker, 2012). Addressing these patterns of political ageism requires sustained advocacy and education efforts that challenge ageist assumptions and demonstrate the value of age inclusion for achieving broader social and environmental goals.

Social resistance to age-inclusive low-carbon energy transitions may also come from younger generations who may view age-related concerns as competing with environmental priorities or who may not understand the importance of age inclusion for achieving effective and sustainable energy transitions (Jenkins et al., 2018). Building intergenerational support for age-just, low-carbon energy transitions requires developing communication and engagement strategies that demonstrate how age inclusion can support rather than compete with environmental goals while also promoting intergenerational solidarity and cooperation.

8.2.5 Coordination and Governance Challenges

The implementation of age-just, low-carbon energy transitions also faces significant coordination and governance challenges that reflect the complex and multidimensional nature of energy ageism and the need for coordinated responses across multiple levels of government and multiple policy domains (Żuk and Żuk, 2024a). These coordination challenges are particularly pronounced because energy ageism intersects with multiple policy areas, including energy, aging, housing, healthcare, and social services, requiring coordination between organizations and agencies that may have different priorities, cultures, and approaches to policy implementation.

Coordination challenges may be particularly pronounced at the local level, where multiple agencies and organizations may be involved in delivering services to older adults but may not have established mechanisms for coordinating their efforts around energy-related issues (McCauley et al., 2013). For example, aging services agencies may not have expertise in energy issues, while energy agencies may not have expertise in aging issues, creating gaps in service delivery that may leave older adults without adequate support for navigating energy transitions.

Addressing coordination challenges requires developing new governance mechanisms and institutional arrangements that can facilitate coordination between different organizations and agencies while also ensuring that age-related concerns are adequately addressed in energy policy and program implementation (Heffron, 2022). This might involve creating inter-agency coordination committees, developing integrated service delivery models, and establishing clear roles and responsibilities for different organizations in addressing energy ageism.

Coordination challenges may also be pronounced at the regional and national levels, where energy policy, social policy, and aging policy may be developed and implemented by different agencies with different priorities and approaches (Upham et al., 2022). Addressing these challenges requires high-level political commitment to age inclusion and the development of policy frameworks that integrate age considerations across different policy domains while also ensuring adequate resources and accountability mechanisms for implementation.

The temporal dimensions of coordination challenges are also important, as low-carbon energy transitions occur over long time periods that may span multiple political cycles and changes in government priorities (Jenkins et al., 2016). Ensuring sustained attention to age inclusion in low-carbon energy transitions requires developing institutional mechanisms that can maintain focus on age-related concerns over time while also adapting to changing circumstances and priorities.

8.2.6 *Measurement and Evaluation Challenges*

The implementation of age-just, low-carbon energy transitions also faces significant challenges related to measurement and evaluation, including the difficulty of developing appropriate indicators and metrics for assessing progress on age inclusion in low-carbon energy transitions and the lack of adequate data systems for monitoring age-related energy outcomes (Jenkins et al., 2016). These measurement and evaluation challenges are important because they affect the ability to assess the effectiveness of age-inclusive energy policies and programs and to make necessary adjustments to improve outcomes.

Traditional energy system indicators and metrics may not adequately capture age-related concerns or may not be disaggregated by age in ways that allow for assessment of age-related impacts (Heffron, 2022). For example, energy burden indicators may not consider the specific circumstances of older adults, energy efficiency program participation rates may not be disaggregated by age, and customer satisfaction surveys may not include questions that are relevant to older adults' experiences.

Developing appropriate indicators and metrics for age-just energy transitions requires understanding the multiple dimensions of energy ageism and the complex ways that energy systems affect older adults' quality of life and well-being (Żuk and Żuk, 2024a). This might involve developing new indicators that capture older adults' experiences of energy systems, disaggregating existing indicators by age and other demographic characteristics, and conducting regular surveys and assessments that focus specifically on older adults' energy experiences.

Data collection challenges may also be significant, as older adults may be underrepresented in traditional energy system data collection efforts or may face barriers to participating in surveys and assessments (Bouzarovski et al., 2023). Addressing these data collection challenges requires developing age-friendly data collection methods and ensuring that older adults are adequately represented in energy system monitoring and evaluation efforts.

Evaluation challenges may also include the difficulty of attributing outcomes to specific policies or programs, given the complex and multidimensional nature of energy ageism and the multiple factors that affect older adults' energy experiences (Upham et al., 2022). This requires developing evaluation approaches that can address complex causation and that can assess the cumulative impacts of multiple interventions over time.

8.3 Future Research Directions and Methodological Innovations

The field of age-just, low-carbon energy transitions would benefit significantly from increased interdisciplinary collaboration that brings together insights from aging studies, energy research, environmental science, public policy, urban planning, and other relevant disciplines (McCauley et al., 2019). The complex and multidimensional nature of energy ageism requires research approaches that can address the interactions between biological, psychological, social, economic, and environmental factors that shape older adults' experiences of energy transitions while also considering how these factors operate across different scales and contexts.

Future research should explore the development of integrated theoretical frameworks that can address the multiple dimensions of energy ageism simultaneously while also considering how these dimensions interact with other forms of inequality and disadvantage (Ciplet, 2021). This might involve developing new conceptual models that integrate insights from aging studies, energy justice research, environmental sociology, urban planning, and other relevant fields while also drawing on interdisciplinary approaches to understanding complex social-environmental systems.

Methodological innovations are also needed to address the challenges of studying energy ageism across different scales and contexts, including the development of new data collection methods that can capture the experiences of diverse groups of older adults, new analytical techniques that can address the complex interactions between different factors, and new evaluation methods that can assess the effectiveness of interventions designed to promote age-inclusive energy transitions (Jenkins et al., 2016). These methodological innovations should draw on advances in mixed-methods research, participatory research, and community-based research that can address the complex and multidimensional nature of energy ageism.

Interdisciplinary research approaches should also address the temporal dimensions of energy ageism and energy transitions, including how energy ageism develops and changes over time, how energy transitions affect different cohorts of older adults differently, and how the cumulative effects of energy-related disadvantages affect older adults' long-term outcomes (Martinson and Berridge, 2015). This requires longitudinal research approaches that can track changes over time while also considering how historical experiences and cohort effects shape older adults' current energy experiences.

The spatial dimensions of energy ageism also require interdisciplinary approaches that can address the complex interactions between built environment factors, social factors, and individual factors that shape older adults' energy experiences (Żuk and Żuk, 2024a). This might involve collaboration between energy researchers, urban planners, geographers, and aging researchers to develop new approaches to understanding how spatial factors affect energy ageism and how spatial interventions can promote age-inclusive energy transitions.

Future research should include more longitudinal studies that can track changes in older adults' experiences of energy transitions over time and identify the factors that contribute to successful adaptation or continued disadvantage (Upham et al., 2022). Longitudinal research is particularly important for understanding how energy ageism develops and changes over time and for evaluating the long-term effectiveness of interventions designed to promote age inclusion in energy transitions.

Longitudinal studies should address multiple dimensions of older adults' energy experiences, including changes in energy costs and affordability, changes in energy technology adoption and use, changes in energy-related health and well-being outcomes, and changes in participation in energy-related decision-making and community activities (Szulecki et al., 2024). These studies should also consider how changes in older adults' circumstances, such as changes in health status, income, or housing situation, affect their energy experiences and their ability to participate in energy transitions.

Comparative studies across different national, regional, and local contexts are also needed to understand how different policy approaches, institutional arrangements, and cultural contexts affect older adults' experiences of energy transitions (Szulecki et al., 2024). Such studies can help identify best practices and successful models that can be adapted to different contexts while also highlighting the importance of context-specific approaches to addressing energy ageism.

Comparative research should address multiple levels of analysis, including comparisons between different countries with different welfare state models and energy system characteristics, comparisons between different regions within countries with different demographic profiles and energy infrastructure, and comparisons between different communities with different approaches to energy planning and service delivery (Heffron, 2022). These comparative studies should also consider how different cultural contexts and social attitudes toward aging affect the manifestation and impact of energy ageism.

International comparative research should also address the global dimensions of energy transitions and aging, including how different stages of economic development, different energy resource endowments, and different demographic transitions affect the relationship between aging and energy transitions (Creti and Ftiti, 2024). This is particularly important given that population aging is a global phenomenon that affects countries at different stages of development and with different energy system characteristics.

Future studies should also include more participatory and community-based approaches that involve older adults as partners in research design, data collection, and analysis rather than simply as subjects of study (McCauley et al., 2013). Participatory research approaches can help ensure that research addresses the questions and concerns that are most important to older adults while also building capacity for community-based solutions to energy ageism and promoting older adults' agency and empowerment in addressing energy-related challenges.

Community-based research approaches can also help identify local solutions and innovations that may not be visible through traditional research methods while also building relationships and trust that can support the implementation of research findings (Fuller and McCauley, 2016). Such approaches are particularly important for understanding the community and social capital factors that affect older adults' experiences of energy transitions and for developing interventions that are responsive to local contexts and needs.

Participatory research approaches should also address the diversity of older adult experiences and ensure that research includes the perspectives of older adults from different backgrounds, including different income levels, racial and ethnic groups, geographic locations, and health statuses (Krekula et al., 2018). This requires attention to how different forms of inequality intersect with age to create different patterns of energy disadvantage and different needs for research and intervention.

Community-based participatory research approaches should also address the power dynamics that may affect older adults' participation in research and ensure that older adults have meaningful control over research processes and outcomes (Calasanti, 2015). This might involve training older adults as co-researchers, providing compensation for participation, and ensuring that research findings are communicated back to participating communities in accessible and useful formats.

Participatory research approaches should also consider how to engage older adults who may face barriers to traditional forms of participation, including those with mobility limitations, communication challenges, or social isolation (Ayalon and Tesch-Römer, 2018). This might involve developing new methods for remote participation, using peer networks to reach isolated older adults, and providing appropriate support and accommodations for participation.

Future research should also address the role of technology and innovation in promoting age-inclusive energy transitions, including research on the development of age-friendly energy technologies, the design of accessible digital-energy services, and the evaluation of technology interventions designed to address energy ageism (Bouzarovski et al., 2023). This research should address both the technical aspects of technology development and the social aspects of technology adoption and use among older adults.

Research on age-friendly energy technology development should address the specific needs and preferences of older adults while also considering how technologies can be designed to accommodate age-related changes in vision, hearing, mobility, and cognitive processing (Furszyfer Del Rio et al., 2021). This might involve user-centered design approaches that involve older adults in technology development processes, usability testing with older adult populations, and evaluation of technology interventions in real-world settings.

Research on digital-energy services should address the factors that affect older adults' adoption and use of digital technologies, including factors related to access, skills, confidence, and trust (Szulecki et al., 2024). This research should also address how digital-energy services can be designed to be more accessible and usable for older adults while also maintaining functionality and efficiency for other user groups.

Innovation research should also address social and community-based innovations that may be particularly beneficial for older adults, including community-energy programs, peer support networks, and social enterprises that focus on addressing energy-related challenges for older adults (McCauley et al., 2013). This research should address how these social innovations can be developed, implemented, and scaled while also considering their effectiveness in addressing different dimensions of energy ageism.

Technology research should also address the ethical and social implications of new energy technologies for older adults, including issues related to privacy, autonomy, and social inclusion (Furszyfer Del Rio et al., 2021). This research should consider how new technologies can be developed and implemented in ways that respect older adults' rights and preferences while also promoting their participation in energy transitions.

Future studies should also address policy and governance questions related to age-inclusive energy transitions, including research on the effectiveness of different policy approaches, the design of governance mechanisms that can promote age inclusion, and the evaluation of policy interventions designed to address energy ageism (Jenkins et al., 2016). This research should address multiple levels of governance, from local to international, and should consider how different governance approaches can promote or hinder age inclusion in energy transitions.

Policy research should address the effectiveness of different types of interventions designed to promote age inclusion in energy transitions, including targeted programs for older adults, universal programs that benefit all age groups, and regulatory approaches that require age inclusion in energy planning and service delivery (Heffron, 2022). This research should use rigorous evaluation methods to assess the impacts of different policy approaches while also considering their cost-effectiveness and feasibility for implementation in different contexts.

Governance research should address the institutional and organizational factors that affect the implementation of age-inclusive energy policies, including the roles of different levels of government, the coordination between different agencies and organizations, and the participation of older adults in governance processes (Upham et al., 2022). This research should also address how governance systems can be reformed to better address age-related concerns while also promoting broader energy transition goals.

Policy research should also address the political economy of age-inclusive energy transitions, including the factors that promote or hinder political support for age inclusion, the role of different stakeholder groups in promoting or opposing age-inclusive policies, and the strategies that can be used to build political coalitions for age-just energy transitions (Fuller and McCauley, 2016). This research should consider how political and economic factors interact to shape policy outcomes while also identifying opportunities for policy change and reform.

International policy research should also address how different countries and regions can learn from each other's experiences with age-inclusive energy transitions, including the factors that affect the transferability of policy approaches across different contexts and the role of international organizations in promoting age inclusion in energy transitions (Creti and Ftiti, 2024). This research should consider how international cooperation can support the development of age-inclusive energy policies while also respecting national sovereignty and local contexts.

Future research should also address the health and well-being dimensions of energy ageism and age-inclusive energy transitions, including research on how energy systems affect older adults' physical and mental health, how energy transitions can be designed to promote health and well-being outcomes, and how health considerations can be integrated into energy planning and policy development (Żuk and Żuk, 2024a). This research should draw on insights from public health, environmental health, and aging research while also considering the social determinants of health that affect older adults' energy experiences.

Health research should address the direct health impacts of energy systems on older adults, including the effects of air pollution from fossil fuel combustion, the health impacts of energy poverty and inadequate heating or cooling, and the health benefits of clean energy transitions (Żuk and Żuk, 2024a). This research should also address how energy transitions can be designed to maximize health benefits while minimizing health risks for older adults.

Mental health research should address the psychosocial impacts of energy system changes on older adults, including the effects of energy insecurity, the stress associated with navigating complex energy systems, and the social isolation that may result from energy-related barriers to social participation (Szulecki et al., 2024). This research should also address how energy transitions can be designed to promote mental health and social well-being outcomes for older adults.

Health equity research should address how energy-related health impacts are distributed across different groups of older adults, including how income, race, geography, and other factors affect older adults' exposure to energy-related health risks and their access to energy-related health benefits (Krekula et al., 2018). This research should also address how energy policies can be designed to promote health equity while also achieving broader energy transition goals.

Research on healthy aging should also address how energy systems can support older adults' ability to age in place and maintain independence, including how energy-efficient and accessible housing can support aging in place, how reliable and affordable energy services can support independent living, and how community-energy programs can support social connection and community engagement (Rowe and Kahn, 1997). This research should consider how energy systems can be designed to support successful aging while also promoting environmental sustainability.

8.4 Policy Implications

The policy implications of this work are far-reaching, suggesting the need for fundamental changes in how energy systems are designed, regulated, and operated to ensure age inclusion across multiple dimensions of energy access, participation, and benefit distribution. This includes not only targeted programs and services for older adults but also broader changes in energy governance frameworks, regulatory approaches, and institutional practices that can help prevent and address energy ageism while also leveraging older adults' capabilities and contributions to energy system transformation.

The work demonstrates that age-inclusive energy transitions require coordinated policy responses across multiple levels of government and multiple policy domains, with effective responses requiring coordination between national energy policy, regional planning, local housing policy, and social services. This coordination is essential because energy ageism operates through complex interactions between different policy domains and scales of governance that cannot be addressed through isolated interventions in any single domain.

The implementation challenges identified in this chapter highlight the need for sustained attention to capacity building, institutional reform, and political coalition building to overcome the barriers that currently limit the development and implementation of age-inclusive energy policies. These challenges include institutional and organizational barriers that reflect broader patterns of ageism, digital divides that may exclude older adults from increasingly digitized energy systems, financial constraints that limit older adults' ability to participate in energy transitions, and political resistance that may prioritize other concerns over age inclusion.

Also, this work demonstrates that age-just low-carbon transitions are not only a matter of social equity but also a practical necessity for achieving successful energy system transformations. Older adults represent a significant and growing proportion of energy consumers in many countries, and their participation and support are essential for achieving the scale and pace of change necessary to address climate change while maintaining social cohesion and democratic legitimacy. Energy transitions that exclude or disadvantage older adults are likely to face greater resistance, achieve lower levels of participation, and ultimately be less effective in achieving their environmental and social goals.

The study also reveals significant opportunities for future research, policy development, and technological innovation that can advance age-just energy transitions through interdisciplinary collaboration, methodological innovation, and sustained attention to the complex and multidimensional nature of energy ageism. The interdisciplinary nature of this field creates opportunities for collaboration across traditional disciplinary boundaries and for the development of new approaches that can address the complex interactions between aging, energy systems, and social justice.

The ultimate goal of age-just, low-carbon energy transitions is not simply to avoid harming older adults in the process of energy system transformation but to actively engage them as partners and beneficiaries in creating more sustainable, equitable, and inclusive energy systems that serve the needs of all members of society. Achieving this goal will require sustained effort across multiple domains—research, policy, technology, and practice—but the potential benefits extend far beyond older adults themselves to encompass more effective and sustainable energy transitions for all members of society.

The integration of age considerations into energy transition planning and implementation represents both a moral imperative and a practical necessity for achieving successful energy system transformation in aging societies. As this review has demonstrated, addressing energy ageism requires not only technical and economic innovations but also social and institutional innovations that can address the complex challenges of ensuring that energy system transformations benefit all members of society while building the broad social support necessary for achieving ambitious climate goals in democratic societies.

The development of age-just, low-carbon energy transitions also represents an opportunity to advance broader goals of social justice and inclusion while demonstrating that environmental and social goals can be mutually reinforcing rather than competing. By explicitly addressing age dimensions in energy transitions, policymakers and practitioners can develop more inclusive and effective approaches to energy system transformation that serve as models for addressing other dimensions of social inequality and exclusion in the context of environmental challenges.

References

Axon, S.; Morrissey, J. Just energy transitions? Social inequities, vulnerabilities and unintended consequences, Build. Cities, 2020, 1(1), pp. 393–411.

Ayalon, L.; Tesch-Römer, C. Contemporary perspectives on ageism. Springer, 2018. https://doi.org/10.1007/978-3-319-73820-8.

Awaworyi Churchill, S.; Smyth, R. Energy poverty and health... Energy Economics, Energy poverty and health: panel data, evidence from Australia. Energy Economics, 2021, 97, 105219.

Bouzarovski, S.; Fuller, S.; Reames, T.G. Handbook on Energy Justice. Cheltenham, UK: Edward Elgar Publishing, 2023. ISBN: 978 1 83910 295 0.

Bouzarovski, S.; Frankowski, J.; Tirado Herrero, S. Low-Carbon Gentrification: When Climate Change Encounters Residential Displacement. International Journal of Urban and Regional Research, 2020, 42, pp. 845–863.

Calasanti, T. Combating ageism: How successful is successful aging? The Gerontologist, 2015, 56(6), 1093–1101.

Ciplet, D. From energy privilege to energy justice: A framework for embedded sustainable development. Energy Research and Social Science, 2021, 75, 101996.

Creti, A.; Ftiti, Z. Energy, just transition, and sustainability: What's new? Energy Economics, 2024, 139, 107872. https://doi.org/10.1016/j.eneco.2024.107872.

Curtis, S.; Oven, K.; Wistow, J.; Dunn, C.; Dominelli, L. Adaptation to extreme weather events in complex health and social care systems: The example of older people's services in England. Environment and Planning C: Politics and Space, 2017, 36(1), pp. 67–91.

Esping-Andersen, G. The Three Worlds of Welfare Capitalism. Princeton University Press: Princeton, New Jersey, 1990.

Fenger, H.J.M. Welfare Regimes in Central and Eastern Europe: Incorporating Post-Communist Countries in a Welfare Regime Typology. Contemporary Issues and Ideas in Social Sciences. Journal of European Social Policy, 2007, 17(2) pp. 1–30.

Frankowski, J.; Tirado Herrero, S. What is in it for me?A people-centered account of household energy transition co-benefits in Poland, Energy Research and Social Science, 2021, 71, 101787.

Fuller, S.; McCauley, D. Framing energy justice: perspectives from activism and advocacy. Energy Research and Social Science, 2016, 11, pp. 1–8.

Furszyfer Del Rio, D.D.; Sovacool, B.K.; Griffiths, S. Culture, energy and climate sustainability, and Smart Home Technologies: A mixed methods comparison of four countries. Energy and Climate Change, 2021, 2, 100035.

Galvin, R. Inequality and Energy. Academic Press: New York, 2020, p. 360.

Goddard, G.; Farrelly, M.A. Just transition management: Balancing just outcomes with just processes in Australian renewable energy transitions. Applied Energy, 2020, 225, pp. 110–123.

Healy, N.; Barry, J. Politicizing energy justice and energy system transitions. Energy Policy, 2017, 107, pp. 451–459.

Heffron, R.J. Applying energy justice into the energy transition. Renewable and Sustainable Energy Reviews, 2022, 156, 111936.

ILO. Guidelines for a just transition towards environmentally sustainable economies and societies for all. 2015. Available online: https://www.ilo.org/sites/default/files/wcmsp5/groups/public/%40ed_emp/%40emp_ent/documents/publication/wcms_432859.pdf

Jenkins, K. Setting energy justice apart from the crowd: Lessons from environmental and climate justice. Energy Research and Social Science, 2018, 39, pp. 117–121.

Jenkins, K.; McCauley, D.; Heffron, R.; Stephan, H.; Rehner, R. Energy justice: A conceptual review. Energy Research and Social Science, 2016, 11, pp. 174–182.

Jenkins, K.; Sovacool, B.K.; McCauley, D. Humanizing sociotechnical transitions through energy justice: An ethical framework for global transformative change. Energy Policy, 2018, 117, pp. 66–74.

Karpinska, L.; Smiech, S. Multiple faces of poverty. Exploring housing-costs-induced energy poverty in Central and Eastern Europe, Energy Research and Social Science, 2023, 105, 103273.

Krekula, C.; Nikander, P.; Wilińska, M. Multiple marginalizations based on age: Gendered ageism and beyond. In L. Ayalon and C. Tesch-Römer (Eds.), Contemporary Perspectives on Ageism (pp. 33–50). Cham: Springer, 2018. https://doi.org/10.1007/978-3-319-73820-8_3

Martinson, M.; Berridge, C. Successful aging and its discontents: A systematic review of the social gerontology literature. The Gerontologist, 2015, 55(1), pp. 58–69.

Matczak, P.; Frankowski, J.; Putkowska-Smoter, R.; Wróblewski, M.; Łoś, I.. Tackling (Not Only) Air Pollution: Cross-sectional Tensions behind State-led Energy Retrofit Program in Poland. Society and Natural Resources, 2023, 36, pp. 1–22.

McCauley, D.; Heffron, R.; Stephan, H.; Jenkins, K.. Advancing Energy Justice: The Triumvirate of Tenets. International Energy Law Review, 2013, 32(3), pp. 107–110

McCauley, D.; Ramasar, V.; Heffron, R.J.; Sovacool, B.K.; Mebratu, D.; Mundaca, L. Energy justice in the transition to low carbon energy systems: Exploring key themes in interdisciplinary research. Applied Energy, 2019, 233–234, pp. 916–921.

Rowe, J.W.; Kahn, R.L. Successful aging. The Gerontologist, 1997, 37(4), pp. 433–440.

Schlosberg, D. Defining Environmental Justice: Theories, Movements, and Nature. Oxford University Press: New York, 2007.

Sokołowski, J.; Bouzarovski, S. Decarbonisation of the Polish residential sector between the 1990s and 2021: A case study of policy failures. Energy Policy, 2022, 163, 112848

Sovacool, B.K. Who are the victims of low-carbon transitions? Towards a political ecology of climate change mitigation. Energy Research and Social Science, 2021, 73 (2021) 101916.

Stephens, J.C. Energy democracy: redistributing power to the people through renewable transformation. Environment, 2019, 61(2), pp. 4–13.

Szulecki, K.; Neerland, M.A.; Tomter, H.; Wæringsaasen, C.A.B.; Żuk, P.; Żuk, P. Ageism, welfare, and the energy transition: a comparative analysis of the perceptions among the elderly in Poland and Norway. Energy, Sustainability and Society, 2024, 14, p. 35. https://doi.org/10.1186/s13705-024-00468-x

Tikkakoski, P.; Aulake, M.; Paloniemi, R. Towards just transition: tackling inequity and structural causes of vulnerability in key environment, health and climate related policies in Finland. Environmental Science & Policy, 2024, 156(2024), 103736.

Upham, P.; Sovacool, B.; Ghosh, B. Just transitions for industrial decarbonisation: A framework for innovation, participation, and justice. Renewable and Sustainable Energy Reviews, 2022, 167, 112699.

Walker, A. The new ageism. The Political Quarterly, 2012, 83(4), pp. 812–819. https://doi.org/10.1111/j.1467-923X.2012.02360.x.

Walker, G. Environmental Justice: Concepts, Evidence and Politics. Routledge: London, 2012, p. 272.

Walker, G.; Day, R. Fuel poverty as injustice. Energy Policy, 2012, 49C, 69–75

Willand, N. Opportunity, ideal or distraction? Exploring stakeholder perceptions of tackling energy poverty and vulnerability among older Australians, Energy Research and Social Science, 2022, 94, 102852.

Żuk, P.; Żuk, P. Energy ageism: The framework of the problem and the challenges of a just energy transition. Environmental Innovation and Societal Transitions, 2022, 43, 237–243.

Żuk, P.; Żuk, P. Social and spatial determinants of energy ageism: Calibrating social policy towards older people under the conditions of energy transition in Polish society. Energy Research and Social Science, 2024a, 118, 103795.

Żuk, P. Žuk, P. Ecology for the rich? Class aspects of the green transition and the threat of right-wing populism as a reaction to its costs in Poland. Sustainability: Science, Practice, and Policy, 2024b, 20(1), 2351231.

Żuk, P.; Żuk, P. The role of trust, information and legal stability in the development of renewable energy: the analysis of non-economic factors affecting entrepreneurs' investments in green energy in Poland. Environment, Development and Sustainability. 2023, 26. 18499–18534.

CHAPTER 9

Conclusions

The findings presented in this monograph underscore that the transition toward sustainable energy systems must be guided not only by technological and economic imperatives but also by the principles of social equity and justice. The analysis has shown that energy systems are deeply embedded within social structures, where access, affordability, and participation are unequally distributed. Social equity therefore emerges as both a normative goal and an analytical framework for assessing the fairness of energy systems and policies.

The interconnections between energy justice and social equity demonstrate that fairness extends beyond equality of access to encompass the quality, reliability, and safety of energy services. Justice is concerned with equitable sharing of benefits and burdens, while equity emphasizes that all individuals should have access to reliable, affordable, and safe energy necessary for human well-being. These two concepts, while distinct, are mutually reinforcing and together define the ethical foundations of a just energy system.

The evidence further highlights that the energy transition—though essential for addressing climate change—produces complex distributional outcomes. These distributional effects of the energy transition are contingent, multi-level, and policy-dependent. The evidence reviewed demonstrates that the energy transition—when guided by inclusive, equity-oriented governance—can generate extensive positive distributional outcomes across individual, community, and national levels. At the micro level, the adoption of clean and decentralized energy technologies enhances households' well-being through improved health, reduced energy expenditures, and greater energy self-sufficiency. These changes not only alleviate energy poverty but also strengthen gender equality by reducing women's domestic workload and enabling broader participation in economic and educational activities. At the meso level, renewable energy deployment stimulates local economic activity, creates employment opportunities, supports entrepreneurship, and fosters skills development, thereby

I. Siksnelyte-Butkiene, D. Streimikiene, *Social Equity in the Energy Sector*, https://doi.org/10.1007/978-3-032-22224-4_9

revitalizing communities and reinforcing social cohesion. At the macro level, the transition contributes to poverty reduction, job creation, regional convergence, gender equality, innovation, and enhanced energy security, while simultaneously advancing climate and sustainability goals. Taken together, these effects illustrate that the energy transition, when governed through fairness and inclusion, serves as a powerful driver of socioeconomic development and long-term resilience.

The analysis of negative distributional effects demonstrates that the energy transition, while indispensable for achieving climate neutrality, is not inherently equitable or universally beneficial. Empirical evidence reveals that without deliberate attention to fairness, participation, and inclusivity, the transition risks reproducing or even deepening existing inequalities. At the micro level, vulnerable households—especially women and low-income populations—often face persistent barriers to benefiting from renewable energy technologies, experience increased labor burdens, and encounter new forms of economic precarity or displacement. At the meso level, large-scale energy projects may cause land loss, livelihood disruption, and weakened food security, particularly where institutional safeguards and participatory mechanisms are absent. At the macro level, unequal access to finance, uneven policy support, and gendered and regional disparities in employment and income distribution exacerbate structural inequities. Moreover, regressive cost burdens associated with carbon pricing or renewable energy subsidies can heighten income inequality and undermine public support for decarbonization. The transition also entails socioecological challenges such as land-use conflicts, biodiversity loss, and degradation of ecosystem services, which further complicate its social acceptability. Collectively, these findings underline that the energy transition must extend beyond technological and environmental objectives and explicitly integrate justice, recognition, and inclusive governance to prevent new inequalities from emerging.

Integrating the principle of intergenerational equity expands the temporal scope of justice in energy governance. Decisions made today shape the environmental, social, and economic conditions inherited by future generations. Ensuring fairness across time requires policy instruments that value long-term well-being over short-term gains, as reflected in the critical role of social discounting debates. Sustainable and just energy systems must therefore internalize the rights of those not yet born, safeguarding ecological integrity and access to essential energy resources.

Energy poverty, as analyzed in this monograph, remains both a manifestation and a driver of social inequity. Measuring it through multidimensional, equity-informed frameworks enables a more comprehensive understanding of deprivation, capturing not only economic constraints but also social participation, reliability, and environmental sustainability. The integration of equity into measurement practices transforms energy poverty assessment into a practical tool for justice-oriented policymaking, supporting the design of interventions that address both structural and distributional causes of deprivation.

The multidimensional framework presented in this monograph provides a comprehensive approach to measuring energy poverty by integrating economic, social, and environmental dimensions while explicitly incorporating the principles of social equity and sustainability. It reflects the complex and interrelated factors that shape households' access to affordable, reliable, and safe energy, moving beyond conventional expenditure-based or income-centered measures. By embedding equity and justice principles, the framework ensures that measurement captures not only material deprivation but also participation, inclusion, and environmental responsibility. This integrated approach strengthens both analytical and policy capacity, allowing for more targeted, effective, and equitable interventions.

Advancing gender equity in the energy sector represents an essential pillar of a just and inclusive energy transition. As emphasized in this monograph, gender disparities are deeply embedded in structural, institutional, and cultural contexts that shape access to energy, participation in governance, and the distribution of benefits from technological and policy change. Addressing these inequities requires not only expanding women's access to modern, affordable, and sustainable energy services but also strengthening their representation in decision-making, education, and employment across the energy value chain. Embedding gender-responsive policies, data-driven monitoring, and targeted capacity-building transforms women from passive beneficiaries into active agents of energy transformation. By integrating gender equity as a cross-cutting principle of energy governance, the transition toward low-carbon systems becomes more inclusive, legitimate, and sustainable—contributing to broader goals of social justice, economic resilience, and intergenerational fairness.

The emergence of the concept of energy ageism marks a significant theoretical and policy advancement in understanding how age-based discrimination manifests within the context of low-carbon energy transitions. The concept reveals that the barriers faced by older adults in accessing, affording, and participating in modern energy systems are not merely individual challenges but reflect deeply embedded structural, institutional, and cultural inequalities. By identifying five key dimensions—economic, technological, spatial, informational, and cultural—the concept of energy ageism exposes how systemic mechanisms within energy policy and governance perpetuate exclusion, vulnerability, and inequity among older populations. The "vicious circle" of energy ageism further demonstrates how these disadvantages reinforce one another over time, entrenching marginalization and reproducing age-based inequalities across generations. Addressing these dynamics demands an integrated, justice-oriented, and intergenerational policy approach that ensures inclusivity across all stages of the energy transition.

This monograph presents an integrative framework for evidence-based and socially equitable energy governance, synthesizing the theoretical, analytical, and policy dimensions explored. This framework emphasizes that the success of energy sector development depends not only on technological innovation and economic growth but also on the systematic inclusion of fairness, justice, and participation in governance processes. By aligning multidimensional assessment tools, data-driven policy design, and adaptive institutional mechanisms, it ensures that equity becomes an inherent feature of energy systems rather than a corrective afterthought. The framework also highlights the importance of continuous learning, stakeholder engagement, and international cooperation to harmonize local implementation with global sustainability and justice objectives.

The book also demonstrates how the theoretical and analytical foundations of social equity can be translated into practice through an applied example focused on age-inclusive approaches in low-carbon energy transitions. By examining governance models, implementation challenges, and methodological innovations, it illustrates how social equity frameworks can be effectively operationalized within real-world energy systems. This applied perspective not only provides a tangible example of integrating justice and inclusion into the energy transition but also offers broader insights for designing equitable and resilient energy policies.

This monograph has demonstrated that the pursuit of social equity in the energy sector is both an ethical imperative and a practical necessity for achieving sustainable and resilient energy sector development. As the global energy landscape evolves in response to climate, technological, and socioeconomic transformations, equity must be understood not as a peripheral concern but as a structural component of effective energy governance. Fairness in access, affordability, and participation determines not only the social legitimacy of energy policies but also their long-term success in delivering low-carbon and inclusive development outcomes.

The manufacturer's authorised representative in the EU is Springer Nature Customer Service Centre GmbH, Europaplatz 3, 69115 Heidelberg, Germany. If you have any concerns regarding our products, please contact ProductSafety@springernature.com

Printed and bound by CPI Group (UK) Ltd, Croydon, CR0 4YY
21/09/2026
02220011-0014